Autodesk
AutoCAD Architecture 2023
Fundamentals

Elise Moss

AUTODESK
Authorized Developer

PUBLICATIONS

SDC Publications
P.O. Box 1334
Mission, KS 66222
913-262-2664
www.SDCpublications.com
Publisher: Stephen Schroff

ISBN-13: 978-1-63057-526-7
ISBN-10: 1-63057-526-7

Preface

No textbook can cover all the features in any software application. This textbook is meant for beginning users who want to gain a familiarity with the tools and interface of AutoCAD Architecture before they start exploring on their own. By the end of the text, users should feel comfortable enough to create a standard model, and even know how to customize the interface for their own use. Knowledge of basic AutoCAD and its commands is helpful but not required. I do try to explain as I go, but for the sake of brevity I concentrate on the tools specific to and within AutoCAD Architecture.

The files used in this text are accessible from the Internet from the book's page on the publisher's website: www.SDCpublications.com. They are free and available to students and teachers alike.

We value customer input. Please contact us with any comments, questions, or concerns about this text.

Elise Moss
elise_moss@mossdesigns.com

Acknowledgments from Elise Moss

This book would not have been possible without the support of some key Autodesk employees.

The effort and support of the editorial and production staff of SDC Publications is gratefully acknowledged. I especially thank Stephen Schroff for his helpful suggestions regarding the format of this text.

Finally, truly infinite thanks to Ari for his encouragement and his faith.

- Elise Moss

About the Author

Elise Moss has worked for the past thirty years as a mechanical designer in Silicon Valley, primarily creating sheet metal designs. She has written articles for Autodesk's Toplines magazine, AUGI's PaperSpace, DigitalCAD.com and Tenlinks.com. She is President of Moss Designs, creating custom applications and designs for corporate clients. She has taught CAD classes at Laney College, DeAnza College, Silicon Valley College, and for Autodesk resellers. Autodesk has named her as a Faculty of Distinction for the curriculum she has developed for Autodesk products. She holds a baccalaureate degree in Mechanical Engineering from San Jose State.

She is married with three sons. Her husband, Ari, is retired from a distinguished career in software development.

Elise is a third-generation engineer. Her father, Robert Moss, was a metallurgical engineer in the aerospace industry. Her grandfather, Solomon Kupperman, was a civil engineer for the City of Chicago.

She can be contacted via email at elise_moss@mossdesigns.com.

More information about the author and her work can be found on her website at www.mossdesigns.com.

Other books by Elise Moss

Autodesk Revit Architecture 2023 Basics
AutoCAD 2023 Fundamentals

Table of Contents

Notes:

Lesson 1:
Desktop Features

AutoCAD Architecture (ACA) enlists object-oriented process systems (OOPS). That means that ACA uses intelligent objects to create a building. This is similar to using blocks in AutoCAD. Objects in AutoCAD Architecture are blocks on steroids. They have intelligence already embedded into them. A wall is not just a collection of lines. It represents a real wall. It can be constrained, has thickness and material properties, and is automatically included in your building schedule.

AEC is an acronym for Architectural/Engineering/Construction.
BID is an acronym for Building Industrial Design.
BIM is an acronym for Building Information Modeling.
AIA is an acronym for the American Institute of Architects.
MEP is an acronym for Mechanical/Electrical/Plumbing.

The following table describes the key features of objects in AutoCAD Architecture:

Feature Type	Description
AEC Camera	Create perspective views from various camera angles. Create avi files.
AEC Profiles	Create AEC objects using polylines to build doors, windows, etc.
Anchors and Layouts	Define a spatial relationship between objects. Create a layout of anchors on a curve or a grid to set a pattern of anchored objects, such as doors or columns.
Annotation	Set up special arrows, leaders, bar scales.
Ceiling Grids	Create reflected ceiling plans with grid layouts.
Column Grids	Define rectangular and radial building grids with columns and bubbles.
Design Center	Customize your AEC block library.
Display System	Control views for each AEC object.
Doors and Windows	Create custom door and window styles or use standard objects provided with the software.
Elevations and Sections	An elevation is basically a section view of a floor plan.
Layer Manager	Create layer standards based on AIA CAD Standards. Create groups of layers. Manage layers intelligently using Layer Filters.
Masking Blocks	Store a mask using a polyline object and attach to AEC objects to hide graphics.
Model Explorer	View a model and manage the content easily. Attach names to mass elements to assist in design.
Multi-view blocks	Blocks have embedded defined views to allow you to easily change view.

Feature Type	Description
Railings	Create or apply different railing styles to a stair or along a defined path.
Roofs	Create and apply various roof styles.
Floorplate slices	Generate the perimeter geometry of a building.
Spaces and Boundaries	Spaces and boundaries can include floor thickness, room height, and wall thickness.
Stairs	Create and apply various stair types.
Tags and Schedules	Place tags on objects to generate schedules. Schedules will automatically update when tags are modified, added, or deleted.
Template Files	Use templates to save time. Create a template with standard layers, text styles, linetypes, dimension styles, etc.
Walls	Create wall styles to determine material composition. Define end caps to control opening and end conditions. Define wall interference conditions.

AutoCAD Architecture sits on top of AutoCAD. It is helpful for users to have a basic understanding of AutoCAD before moving to AutoCAD Architecture. Users should be familiar with the following:

- AutoCAD toolbars and tab on the ribbons
- Zoom and move around the screen
- Manage blocks
- *Draw* and *Modify* commands
- *Model* and *Paper space* (layout)
- Dimensioning and how to create/modify a dimension style

If you are not familiar with these topics, you can still move forward with learning AutoCAD Architecture, but you may find it helpful to have a good AutoCAD textbook as reference in case you get stuck.

The best way to use AutoCAD Architecture is to start your project with massing tools (outside-in design) or space planning tools (inside-out design) and continue through to construction documentation.

The AEC Project Process Model

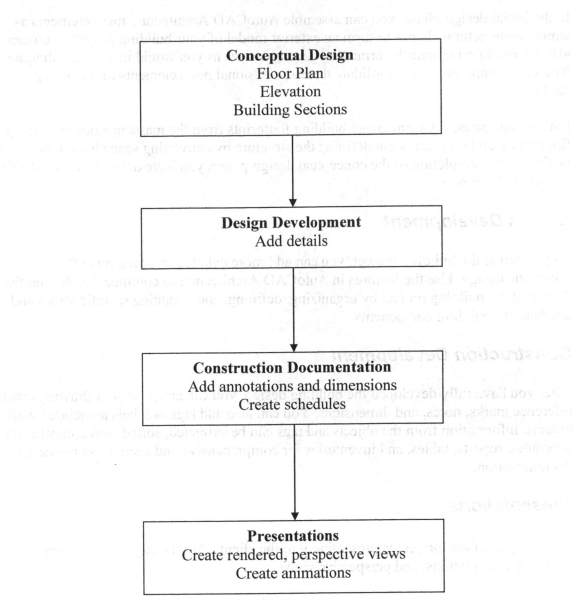

Conceptual Design

In the initial design phase, you can assemble AutoCAD Architecture mass elements as simple architectural shapes to form an exterior model of your building project. You can also lay out interior areas by arranging general spaces as you would in a bubble diagram. You can manipulate and consolidate three-dimensional mass elements into massing studies.

Later in this phase, you can create building footprints from the massing study by slicing floorplates, and you can begin defining the structure by converting space boundaries into walls. At the completion of the conceptual design phase, you have developed a workable schematic floor plan.

Design Development

As you refine the building project, you can add more detailed information to the schematic design. Use the features in AutoCAD Architecture to continue developing the design of the building project by organizing, defining, and assigning specific styles and attributes to building components.

Construction Development

After you have fully developed the building design, you can annotate your drawings with reference marks, notes, and dimensions. You can also add tags or labels associated with objects. Information from the objects and tags can be extracted, sorted, and compiled into schedules, reports, tables, and inventories for comprehensive and accurate construction documentation.

Presentations

A major part of any project is presenting it to the client. At this stage, you develop renderings, animations, and perspective views.

If you use the **QNEW** tool, it will automatically use the template set in the Options dialog.

Express Tools

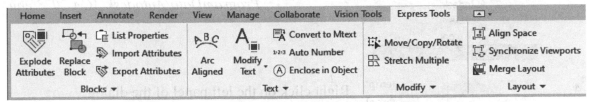

Express Tools are installed with AutoCAD Architecture. If you do not see Express Tools on the menu bar, try typing EXPRESSTOOLS to see if that loads them.

Exercise 1-1:

Setting the Template for QNEW

Drawing Name: New
Estimated Time: 15 minutes

This exercise reinforces the following skills:

- ❑   Use of templates
- ❑   Getting the user familiar with tools and the ACA environment

1.   Launch ACA.

2.   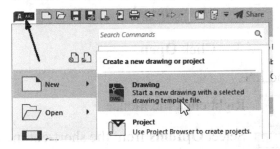   Select the large A to access the Application Menu.

Select **New→Drawing**.

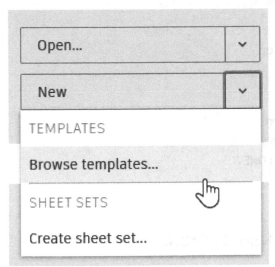

You can also select **New** from the Start tab.

Click the down arrow next to New and select **Browse templates...**

3.	Browse to the *Template* folder under *ProgramData/Autodesk/ACA 2023/enu.*

4.	Right click on the left panel of the dialog.

	Select **Add Current Folder**.

	This adds the template folder to the palette.

5.	*Select the desired template to use.*

	Each template controls units, layers, and plot settings.

	Select *Aec Model (Imperial.Ctb).*

	This template uses Imperial units (feet and inches and a Color Table to control plot settings).

	Click **Open.**

6.	Place your cursor on the command line.

7.	Right click the mouse.

	Select **Options** from the short-cut menu.

8.	Select the **Files** tab.
	Locate the *Template Settings* folder.

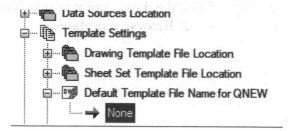

9.	Click on the + symbol to expand.
	Locate the Default Template File Name for QNEW.

10. 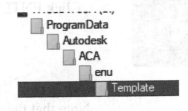 None   Highlight the path and file name listed or None (if no path or file name is listed) and select the **Browse** button.

11. Browse to the *Template* folder or select it on the palette if you added it.

         This should be listed under *Program Data/Autodesk/ ACA 2023/enu*

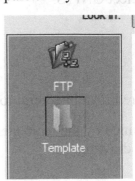

12. Locate the *Aec Model (Imperial Ctb).dwt [Aec Model (Metric Ctb).dwt]* file.

      Click **Open**.

Note that you can also set where you can direct the user for templates. This is useful if you want to create standard templates and place them on a network server for all users to access.

You can also use this setting to help you locate where your template files are located.

13. Click **Apply** and **OK**.

14.   Select the **QNEW** tool button.

15.   Note that you have tabs for all the open drawings.
You can switch between any open drawing files by selecting the folder tab. You can also use the + tab to start a new drawing. The + tab acts the same way as the QNEW button and uses the same default template.

16.   Select the **Model** tab of the open drawing.

17.  Type **units** on the *Command* line.

Notice how you get a selection of commands as you type.

Click ENTER to select UNITS.

18. 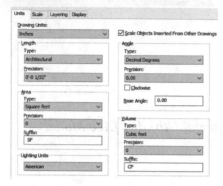 Note that the units are in Architectural format.

The units are determined by the template you set to start your new drawing.

19. Close the drawing without saving.

Templates can be used to preset layers, property set definitions, dimension styles, units, and layouts.

AEC Drawing Options

Menu	Tools→Options
Command line	Options
Context Menu→Options	Place mouse in the graphics area and right click
Shortcut	Place mouse in the command line and right click

Access the Options dialog box.

ACA's Options include five additional AEC specific tabs.
They are AEC Editor, AEC Content, AEC Project Defaults, AEC Object Settings, and
AEC Dimension.

AEC Editor

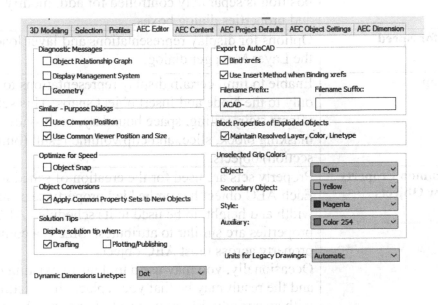

Diagnostic Messages	All diagnostic messages are turned off by default.
Similar-Purpose Dialogs	Options for the position of dialog boxes and viewers.
Use Common Position	Sets one common position on the screen for similar dialog boxes, such as door, wall, and window add or modify dialog boxes. Some dialog boxes, such as those for styles and properties, are always displayed in the center of the screen, regardless of this setting.
Use Common Viewer Position and Sizes	Sets one size and position on the screen for the similar-purpose viewers in AutoCAD Architecture. Viewer position is separately controlled for add, modify, style, and properties dialog boxes.
Optimize for Speed	Options for display representations and layer loading in the Layer Manager dialog.
Object Snap	Enable to limit certain display representations to respond only to the Node and Insert object snaps. This setting affects stair, railing, space boundary, multi-view block, masking block, slice, and clip volume result (building section) objects.
Apply Common Property Sets to New Objects	Property sets are used for the creation of schedule tables. Each AEC object has embedded property sets, such as width and height, to be used in its schedule. AEC properties are similar to attributes. Size is a common property across most AEC objects. Occasionally, you may use a tool on an existing object and the result may be that you replace the existing object with an entirely different object. For example, if you apply the tool properties of a door to an existing window, a new door object will replace the existing window. When enabled, any property sets that were assigned to the existing window will automatically be preserved and applied to the new door provided that the property set definitions make sense.
Solution Tips Problem: The walls are overlapping. Possible Solution: Remove the overlapping wall(s).	Users can set whether they wish a solution tip to appear while they are drafting or plotting. A solution tip identifies a drafting error and suggests a solution.
Dynamic Dimensions Linetype	Set the linetype to be used when creating dynamic dimensions. This makes it easier to distinguish between dynamic and applied dimensions. You may select either continuous or dot linetypes.

Export to AutoCAD Export to AutoCAD ☑ Bind xrefs ☑ Use Insert Method when Binding xrefs Filename Prefix:　　　Filename Suffix: `ACAD-`	Enable Bind Xrefs. If you enable this option, the xref will be inserted as a block and not an xref. Enable Use Insert Method when binding Xrefs if you want all objects from an xref drawing referenced in the file you export to be automatically exploded into the host drawing. If you enable this option, the drawing names of the xref drawings are discarded when the exported drawing is created. In addition, their layers and styles are incorporated into the host drawing. For example, all exploded walls, regardless of their source (host or xref) are located on the same layer. Disable Use Insert Method when binding Xrefs if you want to retain the xref identities, such as layer names, when you export a file to AutoCAD or to a DXF file. For example, the blocks that define walls in the host drawing are located on A-Wall in the exploded drawing. Walls in an attached xref drawing are located on a layer whose name is created from the drawing name and the layer name, such as Drawing1|WallA. Many architects automatically bind and explode their xrefs when sending drawings to customers to protect their intellectual property. Enter a prefix or a suffix to be added to the drawing filename when the drawing is exported to an AutoCAD drawing or a DXF file. In order for any of these options to apply, you have to use the Export to AutoCAD command. This is available under the Files menu.
Unselected Grip Colors	Assign the colors for each type of grip.
Units for Legacy Drawings	Determines the units to be used when opening an AutoCAD drawing in AutoCAD Architecture. 　Automatic – uses the current AutoCAD Architecture units setting 　Imperial – uses Imperial units 　Metric – uses Metric units

Option Settings are applied to your current drawing and saved as the default settings for new drawings. Because AutoCAD Architecture operates in a Multiple Document Interface, each drawing stores the Options Settings used when it was created and last saved. Some users get confused because they open an existing drawing and it will not behave according to the current Options Settings.

AEC Content

| 3D Modeling | Selection | Profiles | AEC Editor | AEC Content | AEC Project Defaults | AEC Object Settings | AEC Dimension |

Architectural / Documentation / Multi-Purpose Object Style Path:

C:\ProgramData\Autodesk\ACA 2020\enu\Styles\Imperial [Browse...]

AEC DesignCenter Content Path:

C:\ProgramData\Autodesk\ACA 2020\enu\AEC Content [Browse...]

Tool Catalog Content Root Path:

C:\ProgramData\Autodesk\ACA 2020\enu\ [Browse...]

IFC Content Path:

C:\ProgramData\Autodesk\ACA 2020\enu\Styles\ [Browse...]

Detail Component Databases: [Add/Remove...]

Keynote Databases: [Add/Remove...]

☑ Display Edit Property Data Dialog During Tag Insertion

Architectural/Documentation/Multipurpose Object Style Path	Allows you to specify a path of source files to be used to import object styles; usually used for system styles like walls, floors, or roofs.
AEC DesignCenter Content Path	Type the path and location of your content files or click Browse to search for the content files.
Tool Catalog Content Root Path	Type the path and location of your Tool Catalog files or click Browse to search for the content files.

IFC Content Path	Type the path and location of your Tool Catalog files or click Browse to search for the content files.
Detail Component Databases	Select the **Add/Remove** button to set the search paths for your detail component files.
Keynote Databases	Select the **Add/Remove** button to set the search paths for your keynote database files.
Display Edit Schedule Data Dialog During Tag Insertion	To attach schedule data to objects when you insert a schedule tag in the drawing, this should be ENABLED.

IFC Content is an object-based building data model that is non-proprietary. It uses open-source code developed by the IAI (International Alliance for Interoperability) to promote data exchange between CAD software. IFC Content might be created in Autodesk Revit, Inventor, or some other software for use in your projects.

Using AEC Content

AutoCAD Architecture uses several pre-defined and user-customizable content including:

- Architectural Display Configurations
- Architectural Profiles of Geometric Shapes
- Wall Styles and Endcap geometry
- Door styles
- Window styles
- Stair styles
- Space styles
- Schedule tables

Standard style content is stored in the AEC templates subdirectory. You can create additional content, import and export styles between drawings.

AEC Object Settings

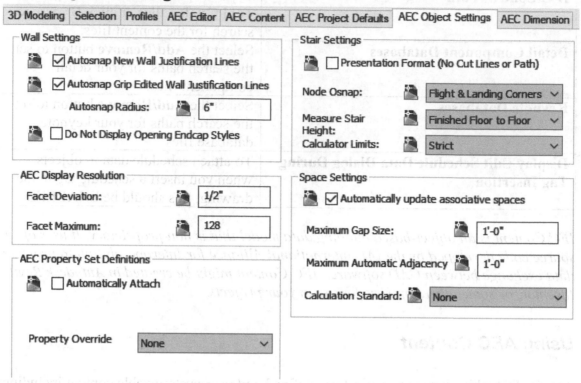

Wall Settings	
Autosnap New Wall Justification Lines	Enable to have the endpoint of a new wall that is drawn within the Autosnap Radius of the baseline of an existing wall automatically snap to that baseline.
	If you select this option and set your Autosnap Radius to 0, then only walls that touch clean up with each other.
Autosnap Grip Edited Wall Justification Lines	Enable to snap the endpoint of a wall that you grip edit within the Autosnap Radius of the baseline of an existing wall.
	If you select this option and set your Autosnap Radius to 0, then only walls that touch clean up with each other.
Autosnap Radius	Enter a value to set the snap tolerance.
Do not display opening end cap styles	Enable Do Not Display Opening Endcap Styles to supClick the display of endcaps applied to openings in walls.
	Enabling this option boosts drawing performance when the drawing contains many complex endcaps.

Stair Settings	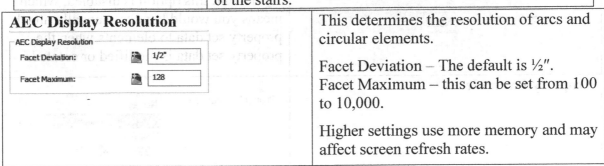
Presentation Format (No Cut Lines or Path)	If this is enabled, a jagged line and directional arrows will not display.

Node Osnap	Determines which snap is enabled when creating stairs:
	    Vertical Alignment
	    Flight and Landing Corners

Measure Stair Height	Rough floor to floor – ignore offsets Finished floor to floor – include top and bottom offsets
Calculator Limits	Strict – stair will display a defect symbol when an edit results in a violation of the Calculation rules. Relaxed – no defect symbol will be displayed when the stair violates the Calculation rules. The Calculation rules determine how many treads are required based on the riser height and the overall height of the stairs.

| AEC Display Resolution

AEC Display Resolution
Facet Deviation: 1/2"
Facet Maximum: 128 | This determines the resolution of arcs and circular elements.

Facet Deviation – The default is ½".
Facet Maximum – this can be set from 100 to 10,000.

Higher settings use more memory and may affect screen refresh rates. |

Space Settings 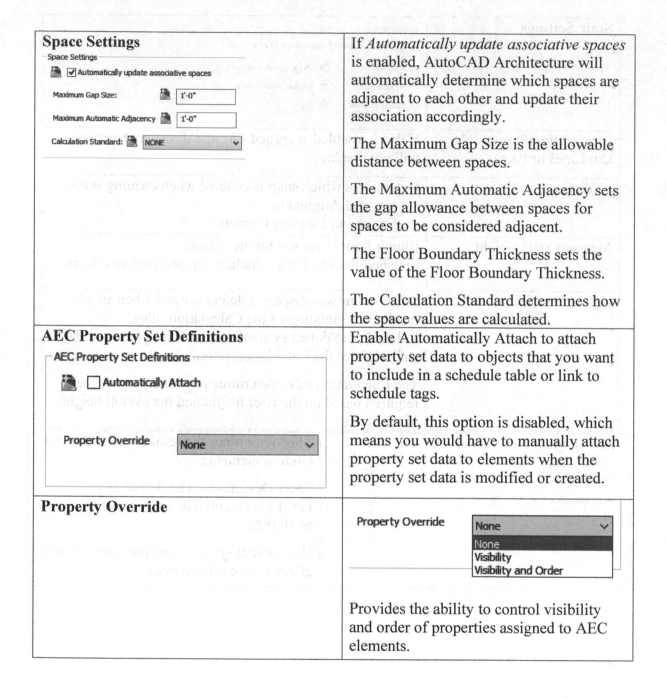	If *Automatically update associative spaces* is enabled, AutoCAD Architecture will automatically determine which spaces are adjacent to each other and update their association accordingly. The Maximum Gap Size is the allowable distance between spaces. The Maximum Automatic Adjacency sets the gap allowance between spaces for spaces to be considered adjacent. The Floor Boundary Thickness sets the value of the Floor Boundary Thickness. The Calculation Standard determines how the space values are calculated.
AEC Property Set Definitions	Enable Automatically Attach to attach property set data to objects that you want to include in a schedule table or link to schedule tags. By default, this option is disabled, which means you would have to manually attach property set data to elements when the property set data is modified or created.
Property Override	Provides the ability to control visibility and order of properties assigned to AEC elements.

AEC Dimension

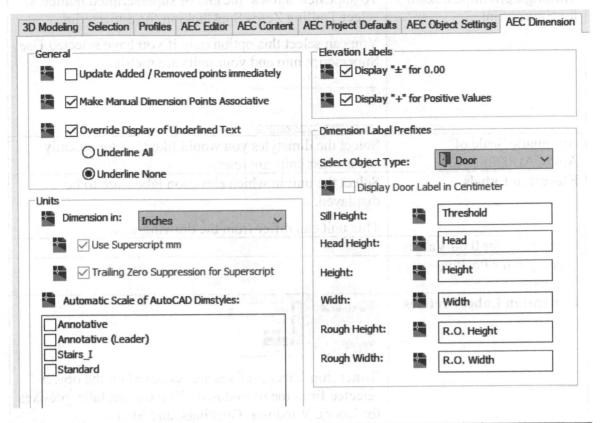

General	
Update Added/Removed points immediately	Enable to update the display every time you add or remove a point from a dimension chain.
Make Manual Dimension Points Associative	Enable if you want the dimension to update when the object is modified.
Override Display of Undefined Text	Enable to automatically underline each manually overridden dimension value.
Underline All	Enable to manually underline overridden dimension values.
Underline None	Enable to not underline any overridden dimension value.
Units	
Dimension in:	Select the desired units from the drop-down list.
Use Superscript mm	If your units are set to meters or centimeters, enable superscripted text to display the millimeters as superscript.

Trailing Zero SupClickion for Superscript	To supClick zeros at the end of superscripted numbers, select Trailing Zeros SupClickion for Superscript. You can select this option only if you have selected Use Superscript mm and your units are metric. 4.12^3
Automatic Scale of AutoCAD Dimstyles	Select the dimstyles you would like to automatically scale when units are reset.
Elevation Labels	Select the unit in which elevation labels are to be displayed. This unit can differ from the drawing unit.
Display +/- for 0.00 Values	
Display + for Positive Values	
Dimension Label Prefixes	Select Object Type: Door / Door / Window / Opening / Stair Display Door Lat Sill Height: Dimension Label Prefixes are set based on the object selected from the drop-down. You can set label prefixes for Doors, Windows, Openings, and Stairs. Select the object, and then set the prefix for each designation. The designations will change based on the object selected.

AEC Project Defaults

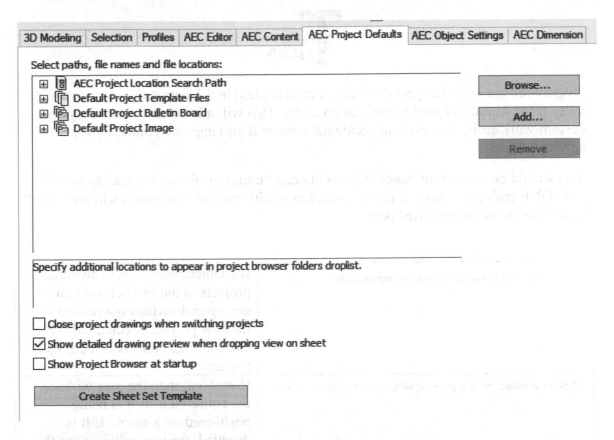

Most BID/AEC firms work in a team environment where they divide a project's tasks among several different users. The AEC Project Defaults allow the CAD Manager or team leader to select a folder on the company's server to locate files, set up template files with the project's title blocks, and even set up a webpage to post project information.

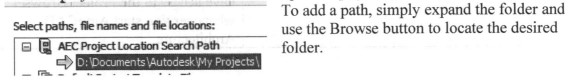

To add a path, simply expand the folder and use the Browse button to locate the desired folder.

You can add more than one path to the AEC Project Location Search Path. All the other folders only allow you to have a single location for template files, the project bulletin board, and the default project image.

I highly recommend that you store any custom content in a separate directory as far away from AutoCAD Architecture as possible. This will allow you to back up your custom work easily and prevent accidental erasure if you upgrade your software application.

You should be aware that AutoCAD Architecture currently allows the user to specify only ONE path for custom content, so drawings with custom commands will only work if they reside in the specified path.

☐ Close project drawings when switching projects	If enabled, when a user switches projects in the Project Navigator, any open drawings not related to the new project selected will be closed. This conserves memory resources.
☑ Show detailed drawing preview when dropping view on sheet	If enabled, then the user will see the entire view as it is being positioned on a sheet. If it is disabled, the user will just see the boundary of the view for the purposes of placing it on a sheet.
☐ Show Project Browser at startup	If enabled, you will automatically have access to the Project Browser whenever you start AutoCAD Architecture.

Architectural Profiles of Geometric Shapes

You can create mass elements to define the shape and configuration of your preliminary study, or mass model. After you create the mass elements you need, you can change their size as necessary to reflect the building design.

- **Mass element**: A single object that has behaviors based on its shape. For example, you can set the width, depth, and height of a box mass element, and the radius and height of a cylinder mass element.

Mass elements are parametric, which allows each of the shapes to have very specific behavior when it comes to the manipulation of each mass element's shape. For example, if the corner grip point of a box is selected and dragged, then the width and depth are modified. It is easy to change the shape to another form by right-clicking on the element and selecting a new shape from the list.

Through Boolean operations (addition, subtraction, intersection), mass elements can be combined into a mass group. The mass group provides a representation of your building during the concept phase of your project.

- **Mass group**: Takes the shape of the mass elements and is placed on a separate layer from the mass elements.

- **Mass model**: A virtual mass object, shaped from mass elements, which defines the basic structure and proportion of your building. A marker appears as a small box in your drawing to which you attach mass elements.

As you continue developing your mass model, you can combine mass elements into mass groups and create complex building shapes through addition, subtraction, or intersection of mass elements. You can still edit individual mass elements attached to a mass group to further refine the building model.

To study alternative design schemes, you can create a number of mass element references. When you change the original of the referenced mass element, all the instances of the mass element references are updated.

The mass model that you create with mass elements and mass groups is a refinement of your original idea that you carry forward into the next phase of the project, in which you change the mass study into floor plates and then into walls. The walls are used to start the design phase.

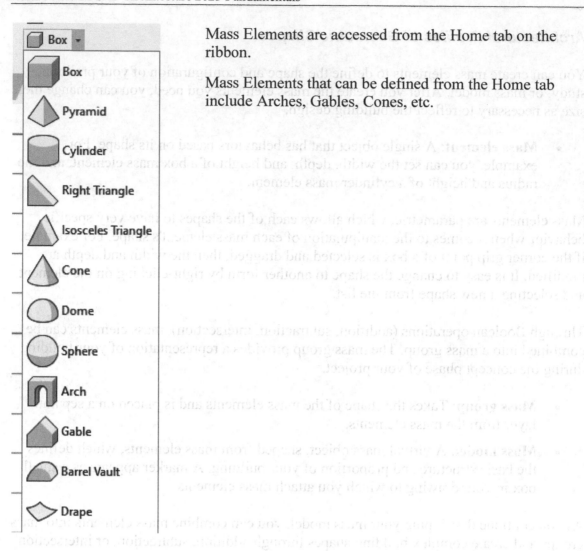

Mass Elements are accessed from the Home tab on the ribbon.

Mass Elements that can be defined from the Home tab include Arches, Gables, Cones, etc.

Exercise 1-2:
Creating a New Geometric Profile

Drawing Name: New
Estimated Time: 15 minutes

This exercise reinforces the following skills:

- ❑ Use of AEC Design Content
- ❑ Use of Mass Elements
- ❑ Use of Views
- ❑ Visual Styles
- ❑ Modify using Properties

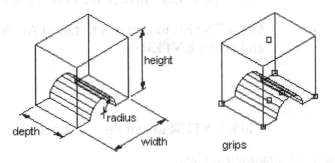

Creating an arch mass element

1. Start a New Drawing.

2. 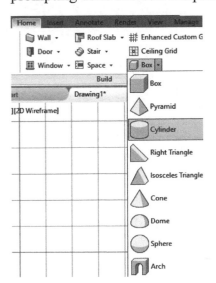 Select the QNEW tool from the Standard toolbar.

3. Because we assigned a template in Exercise 1-1, a drawing file opens without prompting us to select a template.

All the massing tools are available on the **Home** tab on the ribbon in the Build section.

4. 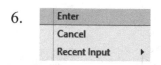   Select the **Arch** tool from the Massing drop down list.

5. Expand the **Dimensions** section on the Properties palette.

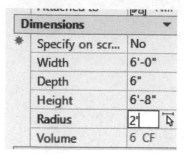

Set Specify on screen to **No**.
Set the Width to **6′ [183]**.
Set the Depth to **6″ [15]**.
Set the Height to **6′ 8″ [203]**.
Set the Radius to **2′ [61]**.

Units in brackets are centimeters.

You need to enter in the ' and " symbols.

6.    Pick a point anywhere in the drawing area.

Click ENTER to accept a Rotation Angle of 0 or right click and select **ENTER**.

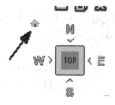

Click **ENTER** to exit the command.

7.   Switch to an isometric view.

To switch to an isometric view, simply go to **View** panel on the Home tab. Under the Views list, select **SW Isometric**.

Or select the Home icon next to the Viewcube.

Our model so far.

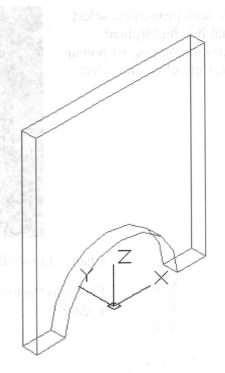

8.

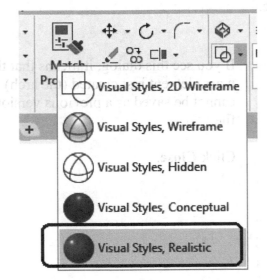

Select **Visual Styles, Realistic** on the View panel on the Home tab to shade the model.

You can also type **SHA** on the command line, then **R** for Realistic.

9. To change the arch properties, select the arch so that it is highlighted. If the properties dialog doesn't pop up automatically, right click and select **Properties**.

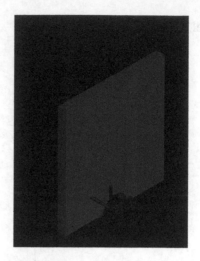

10.

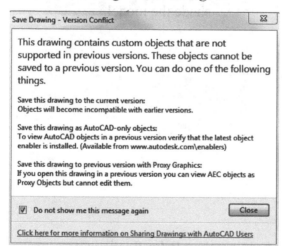

Shadow display	Casts...
Dimensions	▼
Width	6'-0"
Depth	6"
Height	6'-8"
Radius	9"
Volume	20 CF

Change the Radius to **9″** [**15**].

Pick into the graphics window and the arch will update.

11. Click **ESC** to release the grips on the arc.

12. Save the drawing as *ex1-2.dwg*.

Save Drawing - Version Conflict

This drawing contains custom objects that are not supported in previous versions. These objects cannot be saved to a previous version. You can do one of the following things.

Save this drawing to the current version:
Objects will become incompatible with earlier versions.

Save this drawing as AutoCAD-only objects:
To view AutoCAD objects in a previous version verify that the latest object enabler is installed. (Available from www.autodesk.com\enablers)

Save this drawing to previous version with Proxy Graphics:
If you open this drawing in a previous version you can view AEC objects as Proxy Objects but cannot edit them.

☑ Do not show me this message again   Close

Click here for more information on Sharing Drawings with AutoCAD Users

If you see this dialog, it means that the mass element just created (the arch) cannot be saved as a previous version file.

Click **Close**.

The Style Manager

Access the Style Manager from the Manage tab on the ribbon.

The Style Manager is a Microsoft® Windows Explorer-based utility that provides you with a central location in Autodesk AutoCAD Architecture where you can view and work with styles in drawings or from Internet and intranet sites.

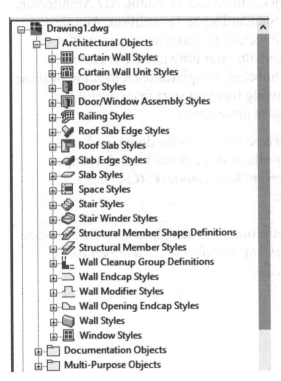

Styles are sets of parameters that you can assign to objects in Autodesk AutoCAD Architecture to determine their appearance or function. For example, a door style in Autodesk AutoCAD Architecture determines what door type, such as single or double, bi-fold or hinged, a door in a drawing represents. You can assign one style to more than one object, and you can modify the style to change all the objects that are assigned that style.

Depending on your design projects, either you or your CAD Manager might want to customize existing styles or create new styles. The Style Manager allows you to easily create, customize, and share styles with other users. With the Style Manager, you can:

- Provide a central point for accessing styles from open drawings and Internet and intranet sites

- Quickly set up new drawings and templates by copying styles from other drawings or templates

- Sort and view the styles in your drawings and templates by drawing or by style type

- Preview an object with a selected style

- Create new styles and edit existing styles

- Delete unused styles from drawings and templates

- Send styles to other Autodesk AutoCAD Architecture users by email

Objects in Autodesk AutoCAD Architecture that use styles include 2D sections and elevations, AEC polygons, curtain walls, curtain wall units, doors, endcaps, railings, roof slab edges, roof slabs, schedule tables, slab edges, slabs, spaces, stairs, structural members, wall modifiers, walls, window assemblies, and windows.

Additionally, layer key styles, schedule data formats, and cleanup group, mask block, multi-view block, profile, and property set definitions are handled by the Style Manager.

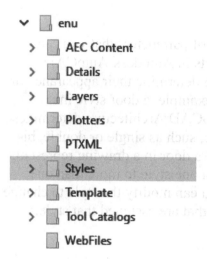

Most of the objects in Autodesk AutoCAD Architecture have a default Standard style. In addition, Autodesk AutoCAD Architecture includes a starter set of styles that you can use with your drawings. The Autodesk AutoCAD Architecture templates contain some of these styles. Any drawing that you start from one of the templates includes these styles.

You can also access the styles for doors, end caps, spaces, stairs, walls, and windows from drawings located in *ProgramData\Autodesk\ACA 2023\enu\Styles*.

Property set definitions and schedule tables are located in the *ProgramData\Autodesk\ACA 2023\enu\AEC Content* folder.

Exercise 1-3:
Creating a New Wall Style

Drawing Name: New using Imperial [Metric] (ctb).dwt
Estimated Time: 30 minutes

This exercise reinforces the following skills:

□   Style Manager
□   Use of Wall Styles

1.    Select the **QNEW** tool from the Quick Access toolbar.

2.   Activate the **Manage** tab on the ribbon.

    Style   Select the **Style Manager** from the **Style & Display** panel.
    Manager

3. The Style Manager is displayed with the current drawing expanded in the tree view. The wall styles in the current drawing are displayed under the wall style type. All other style and definition types are filtered out in the tree view.

    Locate the wall styles in the active drawing.

    Drawing2.dwg
      Architectural Objects
        Curtain Wall Styles
        Curtain Wall Unit Styles
        Door Styles
        Door/Window Assembly Styles
        Railing Styles
        Roof Slab Edge Styles
        Roof Slab Styles
        Slab Edge Styles
        Slab Styles
        Space Styles
        Stair Styles
        Stair Winder Styles
        Structural Member Shape Definitions
        Structural Member Styles
        Wall Cleanup Group Definitions
        Wall Endcap Styles
        Wall Modifier Styles
        Wall Opening Endcap Styles
        Wall Styles
          Standard
        Window Styles

4. Wall St... New   Highlight Wall Styles by left clicking on the name.
    St   Synchror   Right click and select **New**.
    Wind

5. Change the name of the new style by typing **Brick_Block**.

    Wall Styles
      Brick_Block
      Standard

6. In the General tab, type in the description **8″ CMU and 3-1/2″ [200 mm CMU and 90 mm] Brick Wall** in the Description field as shown.

    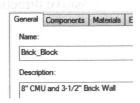
    General | Components | Materials | E
    Name:
    Brick_Block
    Description:
    8″ CMU and 3-1/2″ Brick Wall

7. Select the Components tab.
    Change the Name to **CMU** by typing in the Name field indicated.

Click the drop-down arrow next
to the Edge Offset Button.
Set the Edge Offset to 0.

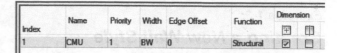

8. A 0 edge offset specifies that the outside edge of the CMU is coincident with the wall baseline.

9. Click the **Width** down arrow button.

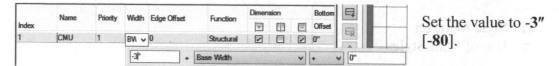

Set the value to **-3″** **[-80]**.

This specifies that the CMU has a fixed width of 3 inches [80 mm] in the negative direction from the wall baseline (to the inside).

You now have one component defined for your wall named CMU with the properties shown.

10. Click the **Add** Component button on the right side of the dialog box.
Next, we'll add wall insulation.

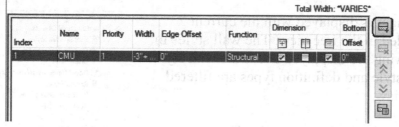

11. 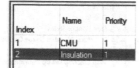 In the Name field, type **Insulation**.

Set the Priority to 1. (The lower the priority number, the higher the priority when creating intersections.)

Set the Edge Offset with a value of **0**.

Set the Component Width as shown, with a value of **1.5″ [38]** and the Base Width set to 0.

This specifies that the insulation has a fixed width of 1.5″ [38 mm] offset in a positive direction from the wall baseline (to the outside).

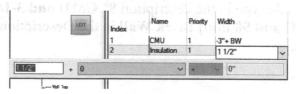

12. 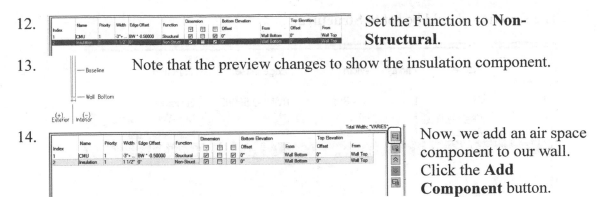 Set the Function to **Non-Structural**.

13. Note that the preview changes to show the insulation component.

14. Now, we add an air space component to our wall. Click the **Add Component** button.

15. Rename the new component **Air Space**.

Set the Edge Offset to **0**.

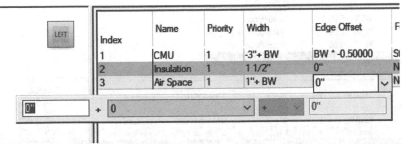

This specifies that the inside edge of the air space is coincident with the outside edge of the insulation.

The width should be set to **1″ [25.4] + Base Width**.

This specifies that the air gap will have a fixed width of 1 inch [25.4 mm] offset in the positive direction from the wall baseline (to the outside).

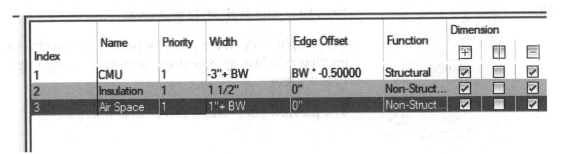

16. Use the **Add Component** tool to add the fourth component.

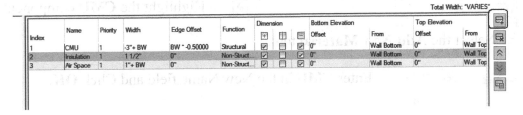

The Name should be **Brick**.
The Priority set to **1**.
The Edge Offset set to **2.5″ [63.5]**.
The Width set to **3.5″ [90]**.

The Function should be set to **Structural**.

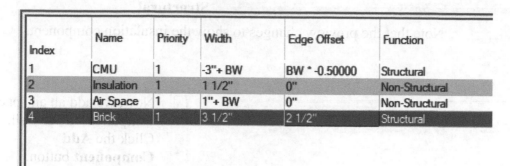

Index	Name	Priority	Width	Edge Offset	Function
1	CMU	1	-3"+ BW	BW * -0.50000	Structural
2	Insulation	1	1 1/2"	0"	Non-Structural
3	Air Space	1	1"+ BW	0"	Non-Structural
4	Brick	1	3 1/2"	2 1/2"	Structural

Index	Name	Priority	Width	Edge Offset	Function
1	CMU	1	-80.0...	0.00	Structural
2	INSULATION	1	38.00...	0.00	Non-Structural
3	AIR SPACE	1	25.40...	0.00	Non-Structural
4	BRICK	1	90.00...	63.50	Structural

Index	Name	Priority	Width	Edge Offset	Function	Dimension ⊞	⫲	⊟	C
1	CMU	1	-3"+ BW	BW * -0.50000	Structural	☑	☑	☑	0
2	Insulation	1	1 1/2"	0"	Non-Structural	☐	☐	☐	0
3	Air Space	1	1"+ BW	0"	Non-Structural	☐	☐	☐	0
4	Brick	1	3 1/2"	2 1/2"	Structural	☑	☑	☑	0

17. The Dimension section allows you to define where the extension lines for your dimension will be placed when measuring the wall. In this case, we are not interested in including the air space or insulation in our linear dimensions. Uncheck those boxes.

18. Note how your wall style previews.

    If you right click in the preview pane, you can change the preview window by zooming, panning, or assigning a Preset View.

    The preview pane uses a DWF-style interface.

19. Select the **Materials** tab.

    Highlight the **CMU** component.

    Select the **Add New Material** tool.

20. Enter **CMU** in the New Name field and Click **OK**.

21. Use the **Add New Material** tool to assign new materials to each component.

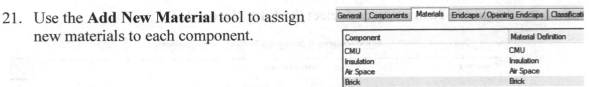

22.

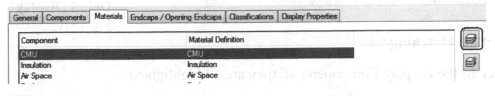

The shrink-wrap material is automatically added. The shrink-wrap material controls the appearance of the wall. Items inside the wall are not shrink-wrapped, while exterior components are – similar to how walls appear in reality.

23. Highlight the **CMU** Component.

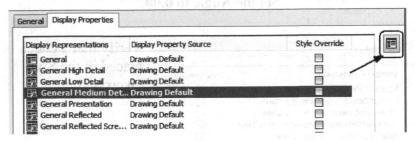

⬛ Select the **Edit Material** tool.

24. Select the **Display Properties** tab.

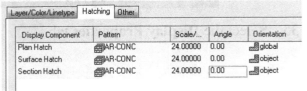

Highlight the **General Medium Detail** Display Representation.

Click on the **Style Override** button.

25. Select the **Hatching** tab.

Highlight **Plan Hatch** and select **Pattern**.

Under Type, select **Predefined**.

Under Pattern Name, select **AR-CONC**.

26. Assign the **AR-CONC** Pattern to all the Display Components.

Set the Angles to **0**.

Display Component	Pattern	Scale/...	Angle	Orientation
Plan Hatch	AR-CONC	24.00000	0.00	global
Surface Hatch	AR-CONC	24.00000	0.00	object
Section Hatch	AR-CONC	24.00000	0.00	object

27. Click **OK** twice to return to the Materials tab of the Wall Properties dialog.

28. Highlight **Insulation** and select the **Edit Material** tool.

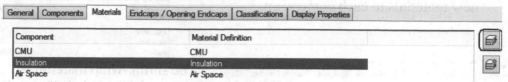

29. Select the **Display Properties** tab.

Click on the **Style Override** button.

Highlight the **General Medium Detail** Display Property.

30. Select the **Hatching** tab.

Select all the Display Components so they are all highlighted.

31.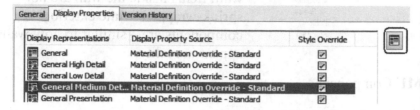

Set the Plan Hatch to the **INSUL** pattern under Predefined type.

Set the Angle to **0.00**.

32. Verify that a check is placed in the Style Override button.

33. Click **OK** to return to the previous dialog.

34. Highlight **Brick** and select the **Edit Material** tool.

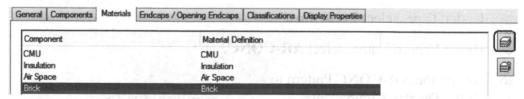

35. Select the **Display Properties** tab.

Highlight the **General Medium Detail** Display Representation.

Click on the **Style**

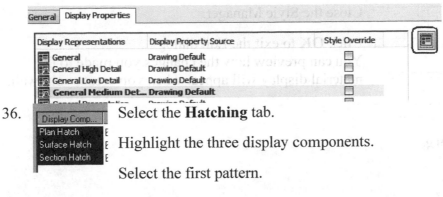

Override button.

36.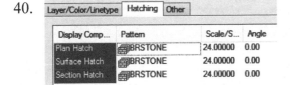

Select the **Hatching** tab.

Highlight the three display components.

Select the first pattern.

37. Set the Type to **Predefined**.

38. Select the **Browse** button.

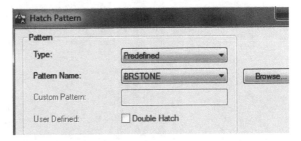

39. Select the **BRSTONE** pattern.

BRSTONE

40. Set the Angle to 0 for all hatches.

Click **OK** twice.

Display Comp...	Pattern	Scale/S...	Angle
Plan Hatch	BRSTONE	24.00000	0.00
Surface Hatch	BRSTONE	24.00000	0.00
Section Hatch	BRSTONE	24.00000	0.00

41.

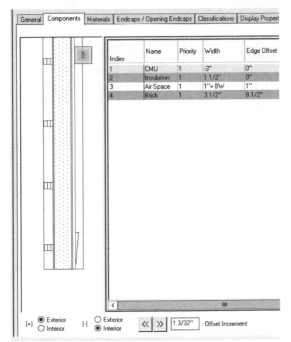

Activate the Components tab. Highlight each component and use the << >> arrows to position the component. By shifting each component position, you can see where it is located. Locate the Brick so it is on the Exterior side and the CMU so it is interior.

☑ Auto Calculate Edge Offset

You can also use the Auto Calculate Edge Offset option to ensure all the components are correct.

42.

Close the Style Manager.

Click **OK** to exit the Edit dialog.

You can preview how the changes you made to the material display will appear when you create your wall.

43. Save as *ex1-3.dwg*.

Exercise 1-4:

Copying a Style to a Template

Drawing Name: Styles.dwg, Imperial (Metric).ctb.dwt
Estimated Time: 20 minutes

This exercise reinforces the following skills:

- Style Manager
- Use of Templates
- Properties dialog

1. 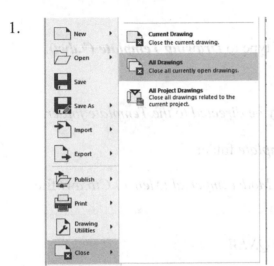 Close any open drawings.

Go to the Application menu and select **Close → All Drawings**.

You can also type CLOSEALL.

2.  Start a New Drawing using plus tab.

3. Activate the **Manage** tab on the ribbon.

Select the **Style Manager** tool.

4. Select the **File Open** tool in the Style Manager dialog.

5.  Locate the *styles.dwg.*
This file is downloaded from the publisher's website.
Click **Open**.

Note: Make sure the Files of type is set to dwg or you will not see your files.

6. 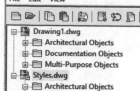 Styles is now listed in the Style Manager.

7. 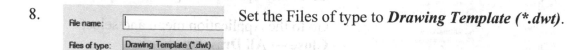 Select the **File Open** tool.

8.  Set the Files of type to ***Drawing Template (*.dwt)***.

You should automatically be directed to the Template folder.

If not, browse to the *Template* folder.

9. 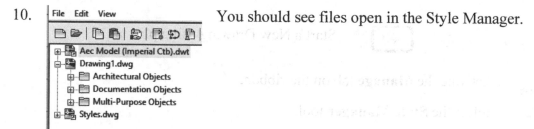 Locate the *AEC Model Imperial (Metric) Ctb.dwt* file. Click **Open**.

This is the default template we selected for QNEW.

10. You should see files open in the Style Manager.

11. 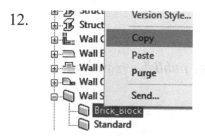 Locate the **Brick_Block** Wall Style in the *Styles.dwg*.

12. Right click and select **Copy**.

13. Locate Wall Styles under the template drawing.

Highlight, right click and select **Paste**.

New
Synchronize with F
Update Standards f
Version Styles...
Copy
Paste

14. You could also just drag and drop the Brick_block style from the Styles file to the template.

Aec Model (Imperial Ctb).dwt
  Architectural Objects
  Documentation Objects
  Multi-Purpose Objects
Drawing3.dwg
  Architectural Objects
  Documentation Objects
  Multi-Purpose Objects
Styles1.dwg
  Architectural Objects
    Curtain Wall Styles
    Curtain Wall Unit Styles
    Door Styles
    Door/Window Assembly Style
    Railing Styles
    Roof Slab Edge Styles
    Roof Slab Styles
    Slab Edge Styles
    Slab Styles
    Space Styles
    Stair Styles
    Stair Winder Styles
    Structural Member Shape Def
    Structural Member Styles
    Wall Cleanup Group Definitio
    Wall Endcap Styles
    Wall Modifier Styles
    Wall Opening Endcap Styles
    Wall Styles
      Brick_Block
      Standard
    Window Styles

15. Apply   Click **Apply**.

16. The following drawing has changed.
C:\ProgramData\Autodesk\ACA
2020\enu\Template\Aec Model (Imperial
Ctb).dwt

Do you wish to save changes to this file?

Yes   No

Click **Yes**.

17. Close the Style Manager.

18.  Look at the drawing folder tabs. The Styles.dwg and the template drawing are not shown. These are only open in the Style Manager.

Drawing1*   X

19. Start   +   Select the **plus** tab to start a new drawing.

20.  Activate the Home tab on the ribbon.
Select the **Wall** tool.

21. 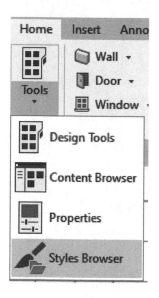 The Properties dialog should pop up when the WallAdd tool is selected.

In the Style drop-down list, note that Brick_Block is available.

If you don't see the wall style in the drop-down list, verify that you set the default template correctly in Options.

22. Click Escape to end the WallAdd command.

23. Close all drawing files without saving.

The Style Browser

 The Styles Browser can be launched from the Home ribbon under the Tools drop-down.

The Styles Browser displays the styles available for various elements.

You can see the styles that are available in the Content Library as well as the active drawings.

You can use the Styles Browser to copy and paste styles between drawings.

Exercise 1-5:

Copying a Style using the Style Browser

Drawing Name: Stylesbrowser.dwg
Estimated Time: 10 minutes

This exercise reinforces the following skills:

- Style Browser
- Use of Styles
- Importing a Style into a Drawing

1.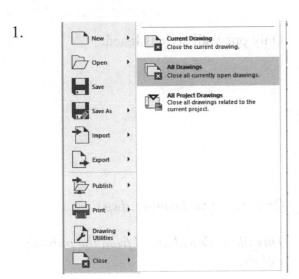

   Close any open drawings.

   Go to the Application menu and select **Close → All Drawings.**

   You can also type CLOSEALL.

2.

   Start a New Drawing using plus tab.

3.

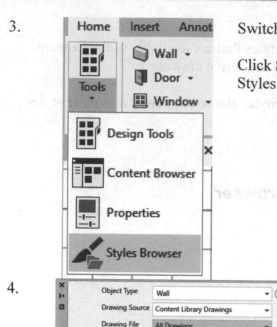

Switch to the Home ribbon.

Click **Styles Browser** under Tools to launch the Styles Browser.

4.

On the Drawing Source field, click the down arrow and select **Current Drawing**.

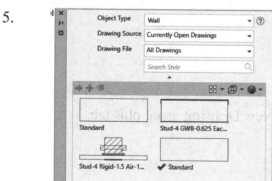

Only one wall style is listed.

5.

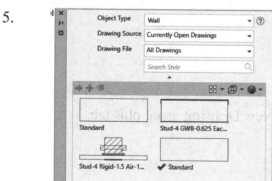

Open the *stylesbrowser.dwg*.

This file is downloaded from the publisher's website.

The Styles Browser shows all the wall styles available in this file.

6. Click on the **Drawing2** tab (the new drawing you created at the start of the exercise).

Notice the Drawing Source has changed to Currently Open Drawings.

The green check indicates any style being used in any of the open drawings.

7. Highlight the Stud-4 Rigid-1.5 Air wall style.

Right click and select **Import styles**.

Note you can hold down the CTL key to select more than one style to be imported.

Close the *stylesbrowser.dwg* without saving.

8. Right click on the Stud-4 Rigid-1.5 Air wall style.

Click **Add object**.

9. Click a point to start placing a wall.
Drag the mouse to the right.
Click to place an end point for the wall.
Click ENTER to exit the command.

Close the Style Browser.

10. Click the **Wall** tool on the Home ribbon.

11. Note that the imported wall style is available in the drawing now.

Place a wall using the imported style.

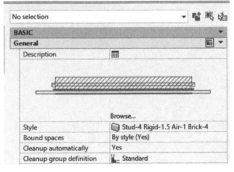

12. Close without saving.

Visual Styles are stored with the drawing.

Visual Styles

Visual styles determine the display of edges, lighting, and shading in each viewport.

You can choose a predefined visual style in the upper-left corner of each viewport. In addition, the Visual Styles Manager displays all styles available in the drawing. You can choose a different visual style or change its settings at any time. ACA comes with several pre-defined visual styles or you can create your own.

Exercise 1-6:

Creating a New Visual Style

Drawing Name: visual_style.dwg
Estimated Time: 15 minutes

This exercise reinforces the following skills:

- ❑ Workspaces
- ❑ Use of Visual Styles
- ❑ Controlling the Display of Objects

1.   Open *visual_style.dwg*.

2.      Activate the **View** tab on the ribbon.

3.   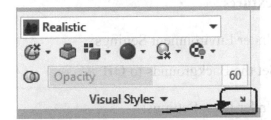   Launch the Visual Styles Manager.

To launch, select the small arrow in the lower right corner of the Visual Styles panel.

4.   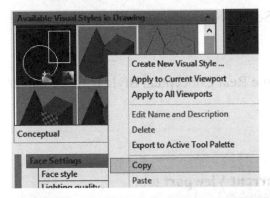   Highlight the **Conceptual** tool.

Right click and select **Copy**.

5.   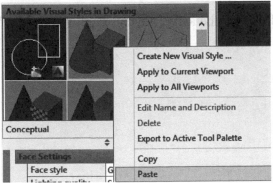   Right click and select **Paste**.

6.

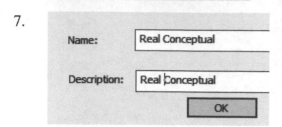

Highlight the copied tool.
Right click and select **Edit Name and Description**.

7.

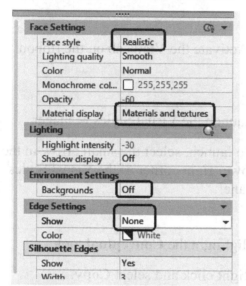

Change the name and description to **Real Conceptual**.

Click **OK**.

8.

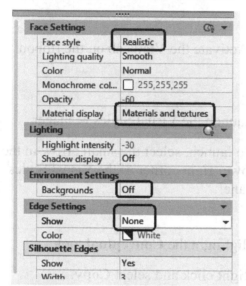

Under Face Settings:

Set the Face Style to **Realistic**.

Set Material display to **Materials and textures**.

Under Environment Settings:

Set the Backgrounds to **Off**.

Under Edge Settings:

Set the Show to **None**.

9.

Verify that the Real Conceptual Style is highlighted/selected.

Right click and select the **Apply to Current Viewport** tool.

[−][SW Isometric][Real Conceptual]

Note that the active visual style is listed in the upper right corner of the drawing.

10.   Save the drawing as *ex1-6.dwg*.

Exercise 1-7:

Creating an Autodesk Cloud Account

Drawing Name: (none, start from scratch)
Estimated Time: 5 Minutes

This exercise reinforces the following skills:

❑   Create an account on Autodesk.

You must have an internet connection to access the Autodesk server. Many government agencies do not allow the use of cloud accounts due to security concerns. Autodesk's cloud account is useful for students, home users, and small shops.

1.   Next to the Help icon at the top of the screen:

Select **Sign In to Autodesk account**.

2.   If you have an existing Autodesk ID, you can use it.

If not, select the link that says **Create Account.**

You may have an existing Autodesk ID if you have registered Autodesk software in the past. Autodesk's cloud services are free to students and educators. For other users, you are provided a small amount of initial storage space for free.

I like my students to have an on-line account because it will automatically back up their work.

3.

Create account

First name

Last name

Email

Confirm email

Password

☐ I agree to the Autodesk **Terms of Use** and acknowledge the **Privacy Statement.**

CREATE ACCOUNT

ALREADY HAVE AN ACCOUNT? **SIGN IN**

Fill in the form to create an Autodesk ID account.

Be sure to write down the ID you select and the password.

4.

Account created

This single account gives you access to all your Autodesk products

☐ Check this box to receive electronic marketing communications from Autodesk on news, trends, events, special offers and research surveys. You can **manage your** preferences or unsubscribe at any time. To learn more, see the **Autodesk Privacy Statement.**

DONE

You should see a confirmation window stating that you now have an account.

5. elise_moss

You will see the name you selected as your Autodesk ID in the Cloud sign-in area.

Tips & Tricks

Use your Autodesk account to create renderings, back up files to the Cloud so you can access your drawings on any internet-connected device, and to share your drawings.

Exercise 1-8:
Tool Palettes

Drawing Name: tool_palettes.dwg
Estimated Time: 15 minutes

This exercise reinforces the following skills:

- Use of AEC Design Content
- Use of Wall Styles

1.  Open the tool_palettes.dwg.

2.   Activate the Home tab on the ribbon.

   Select **Tools→Design Tools** to launch the tool palette.

3.   Right click on the Design Tools palette title bar.

   Click **New Palette**.

   Move
   Size
   Close
   Allow Docking

   Anchor Left <
   Anchor Right >

   Auto-hide
   Transparency...
   View Options...

   New Palette
   Rename Palette Set

4.   Rename the new palette **Walls**.

   Walls

5.

Click the **Design** tab on the tool palette.

Use the scroll bar to scroll to the bottom of the tool palette.

Locate the Content Browser icon.

Click on the **Content Browser** icon to launch.

6.

Design Tool
Catalog -
Imperial

Locate the **Design Tool Catalog – Imperial.**

Click on the **Design Tool Catalog – Imperial** to open.

7.

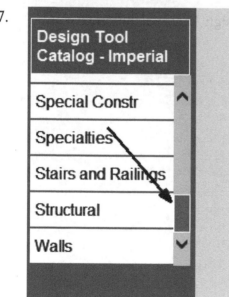

Use the scroll bar to scroll down.

Click on **Walls**.

8.

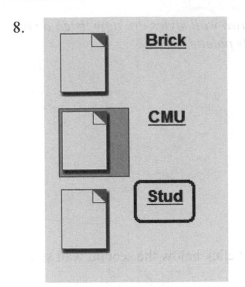

Click on the **Stud** link to look at the stud walls.

9.

Locate the **Stud – 2.5 GWB – 0.625 2 Layers Each Side**.

10.

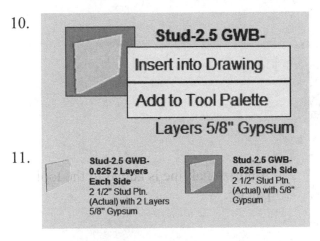

Click on the wall icon.
Select **Add to Tool Palette**.

11.

Locate the **Stud – 2.5 GWB – 0.625 Each Side**.

12.

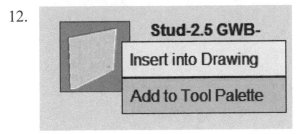

Click on the wall icon.
Select **Add to Tool Palette**.

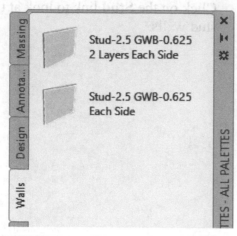

The two wall styles are now listed on the Walls palette.

13.

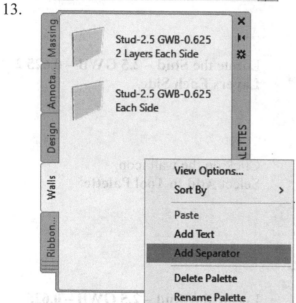

Right click below the second wall style.

Click **Add Separator**.

14.

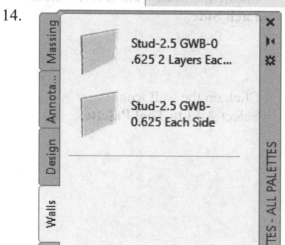

A horizontal line is added to the tool palette.

15.

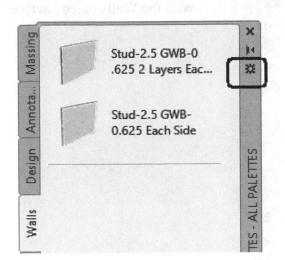

On the Walls tool palette:

Select the **Properties** button.

16.

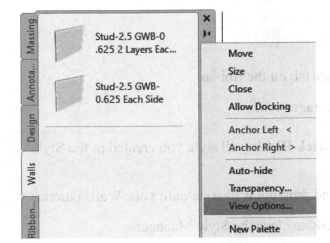

Select **View Options**.

17.

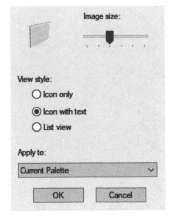

Adjust the Image Size and select the desired View style.

You can apply the preference to the Current Palette (the walls palette only) or all palettes.

Click **OK**.

18.

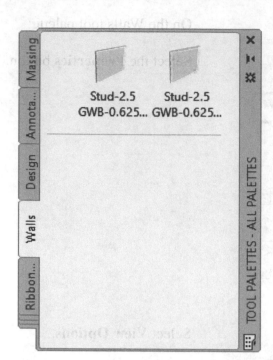

Verify that the tool palette is open with the Walls palette active.

19.   Activate the **Manage** tab on the ribbon.

Select the **Style Manager** tool.

20. 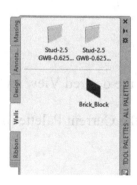 Locate the **Brick-Block** wall style you created in the Style Manager.

21. Drag and drop the wall style onto your Walls palette.

Click **OK** to close the Style Manager.

22.

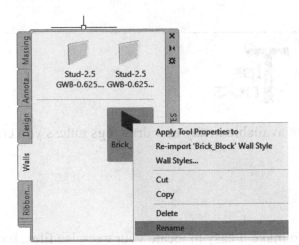

Highlight the **Brick_Block** tool.
Right click and select **Rename**.

23.

Change the name to **8″ CMU- 3.5″ Brick [200 mm CMU – 90 mm]**.

24. Select the new tool and draw a wall.
You may draw a wall simply by picking two points.
If you enable the ORTHO button on the bottom of the screen, your wall will be straight.

25.

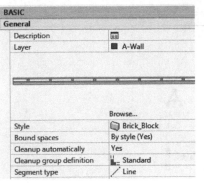

Select the wall you drew.

Right click and select **Properties**.

Note that the wall is placed on the A-Wall Layer and uses the Brick-Block Style you defined.

26. Save your file as *Styles3.dwg*.

The wall style you created will not be available in any new drawings unless you copy it over to the default template.

To make it easy to locate your exercise files, browse to where the files are located when you save the file. Right click on the left pane and select Add Current Folder.

A fast way to close all open files is to type CLOSEALL on the command line. You will be prompted to save any unsaved files.

Layer Manager

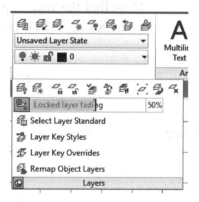

Back in the days of vellum and pencil, drafters would use separate sheets to organize their drawings, so one sheet might have the floor plan, one sheet the site plan, etc. The layers of paper would be placed on top of each other and the transparent quality of the vellum would allow the drafter to see the details on the lower sheets. Different colored pencils would be used to make it easier for the drafter to locate and identify elements of a drawing, such as dimensions, electrical outlets, water lines, etc.

When drafting moved to Computer Aided Design, the concept of sheets was transferred to the use of Layers. Drafters could assign a Layer Name, color, linetype, etc. and then place different elements on the appropriate layer.

AutoCAD Architecture has a Layer Management system to allow the user to implement AIA standards easily.

🔲 *Layer Manager*

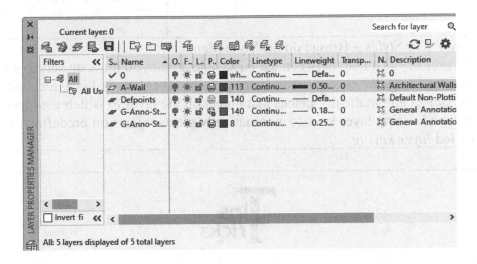

The Layer Manager helps you organize, sort, and group layers, as well as save and coordinate layering schemes. You can also use layering standards with the Layer Manager to better organize the layers in your drawings.

When you open the Layer Manager, all the layers in the current drawing are displayed in the right panel. You can work with individual layers to:

- Change layer properties by selecting the property icons
- Make a layer the current layer
- Create, rename, and delete layers

If you are working with drawings that contain large numbers of layers, you can improve the speed at which the Layer Manager loads layers when you open it by selecting the Layer Manager/Optimize for Speed option in your AEC Editor options.

The Layer Manager has a tool bar as shown.

	Layer Standards You can import Layer Standards from an existing drawing or Export Layer Standards using the current drawing.

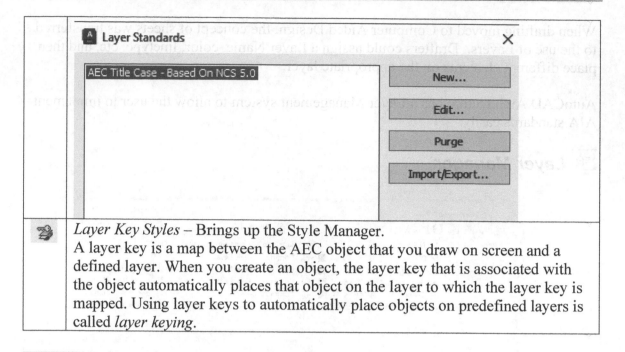

	Layer Key Styles – Brings up the Style Manager. A layer key is a map between the AEC object that you draw on screen and a defined layer. When you create an object, the layer key that is associated with the object automatically places that object on the layer to which the layer key is mapped. Using layer keys to automatically place objects on predefined layers is called *layer keying*.

The layer key styles in Autodesk AutoCAD Architecture, Release 3 and above replace the use of LY files. If you have LY files from S8 or Release 1 of AutoCAD Architecture that you want to use, then you can create a new layer key style from an LY file. On the command line enter **-AecLYImport** to import legacy LY files.

	Layer Key Overrides You can apply overrides to any layer keys within a layer key style that is based on a layer standard. The structure of the layer name of each layer that each layer key maps an object to is determined by the descriptive fields in the layer standard definition. You can override the information in each field according to the values set in the layer standard definition. You can allow overrides on all the layer keys within a layer key style, or you can select individual layer keys that you want to override. You can also choose to allow all of the descriptive fields that make up the layer name to be overridden, or you can specify which descriptive fields you want to override.
	Load Filter Groups This tool allows you to load a saved filter group.
	Save Filter Groups This tool allows you to save a defined filter group, so that you can use it in other drawings.
	New Property Filter
	New Group Filter

	New Standards Filter
	New Layer Creates a new layer.
	New Layer from Standard Creates a new layer based on an existing layer standard. 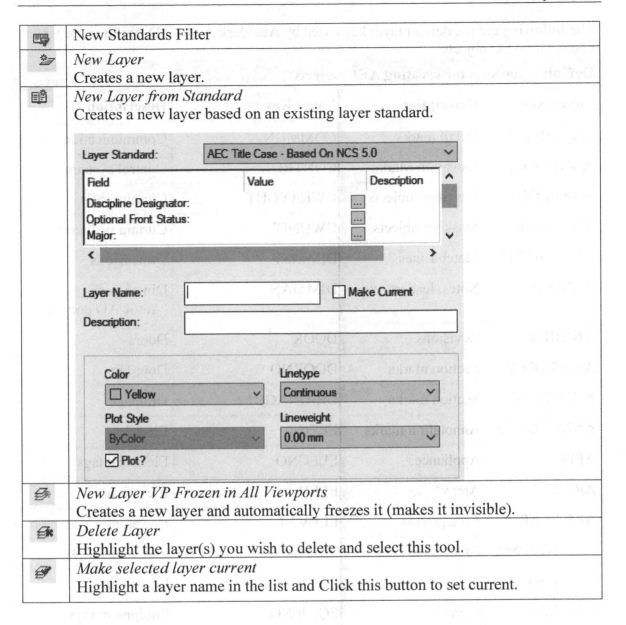
	New Layer VP Frozen in All Viewports Creates a new layer and automatically freezes it (makes it invisible).
	Delete Layer Highlight the layer(s) you wish to delete and select this tool.
	Make selected layer current Highlight a layer name in the list and Click this button to set current.

The following are the default layer keys used by Autodesk AutoCAD Architecture when you create AEC objects.

Default layer keys for creating AEC objects			
Layer Key	**Description**	**Layer Key**	**Description**
ANNDTOBJ	Detail marks	COMMUN	Communication
ANNELKEY	Elevation Marks	CONTROL	Control systems
ANNELOBJ	Elevation objects	CWLAYOUT	Curtain walls
ANNMASK	Masking objects	CWUNIT	Curtain wall units
ANNMATCH	Match Lines	DIMLINE	Dimensions
ANNOBJ	Notes, leaders, etc.	DIMMAN	Dimensions (AutoCAD points)
ANNREV	Revisions	DOOR	Doors
ANNSXKEY	Section marks	DOORNO	Door tags
ANNSXOBJ	Section marks	DRAINAGE	Drainage
ANNSYMOBJ	Annotation marks	ELEC	Electric
APPL	Appliances	ELECNO	Electrical tags
AREA	Areas	ELEV	Elevations
AREAGRP	Area groups	ELEVAT	Elevators
AREAGRPNO	Area group tags	ELEVHIDE	Elevations (2D)
AREANO	Area tags	EQUIP	Equipment
CAMERA	Cameras	EQUIPNO	Equipment tags
CASE	Casework	FINCEIL	Ceiling tags
CASENO	Casework tags	FINE	Details- Fine lines
CEILGRID	Ceiling grids	FINFLOOR	Finish tags
CEILOBJ	Ceiling objects	FIRE	Fire system equip.
CHASE	Chases	FURN	Furniture
COGO	Control Points	FURNNO	Furniture tags
COLUMN	Columns	GRIDBUB	Plan grid bubbles
GRIDLINE	Column grids	SITE	Site

Layer Key	Description	Layer Key	Description
HATCH	Detail-Hatch lines	SLAB	Slabs
HIDDEN	Hidden Lines	SPACEBDRY	Space boundaries
LAYGRID	Layout grids	SPACENO	Space tags
LIGHTCLG	Ceiling lighting	SPACEOBJ	Space objects
LIGHTW	Wall lighting	STAIR	Stairs
MASSELEM	Massing elements	STAIRH	Stair handrails
MASSGRPS	Massing groups	STRUCTBEAM	Structural beams
MASSSLCE	Massing slices	STRUCTBEAMIDEN	Structural beam tags
MED	Medium Lines	STRUCTBRACE	Structural braces
OPENING	Wall openings	STRUCTBRACEIDEN	Structural brace tags
PEOPLE	People	STRUCTCOLS	Structural columns
PFIXT	Plumbing fixtures	STRUCTCOLSIDEN	Structural column tags
PLANTS	Plants - outdoor	SWITCH	Electrical switches
PLANTSI	Plants - indoor	TITTEXT	Border and title block
POLYGON	AEC Polygons	TOILACC	Arch. specialties
POWER	Electrical power	TOILNO	Toilet tags
PRCL	Property Line	UTIL	Site utilities
PRK-SYM	Parking symbols	VEHICLES	Vehicles
ROOF	Rooflines	WALL	Walls
ROOFSLAB	Roof slabs	WALLFIRE	Fire wall patterning
ROOMNO	Room tags	WALLNO	Wall tags
SCHEDOBJ	Schedule tables	WIND	Windows
SEATNO	Seating tags	WINDASSEM	Window assemblies
SECT	Miscellaneous sections	WINDNO	Window tags
SECTHIDE	Sections (2D)		

Display Manager

The Display Manager is a utility for managing display configurations, display sets, and display representations: renaming them or deleting them, copying them between drawings, emailing them to other users, and purging unused elements from drawings.

The display system in AutoCAD Architecture is designed so that you only have to draw an architectural object once. The appearance of that object then changes automatically to meet the display requirements of different types of drawings, view directions, or levels of detail.

You can change the display settings for an element by using the Display tab on the Properties palette.

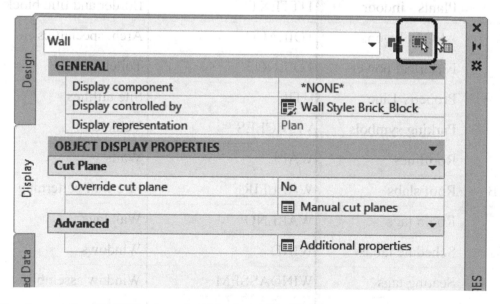

To change the display using the Display tab, click Select Components, select an object display component (like a hatch or a boundary), and then select or enter a new value for the display property you want to change (such as color, visibility, or lineweight).

Exercise 1-9:

Exploring the Display Manager

Drawing Name: display_manager.dwg
Estimated Time: 15 minutes

This exercise reinforces the following skills:

❑   Display Manager

1.   Open *display_manager.dwg.*

2.   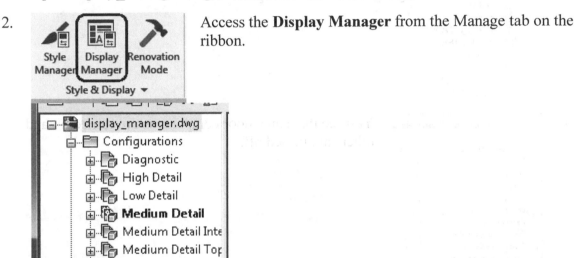   Access the **Display Manager** from the Manage tab on the ribbon.

The display system in Autodesk® AutoCAD Architecture controls how AEC objects are displayed in a designated viewport. By specifying the AEC objects you want to display in a viewport and the direction from which you want to view them, you can produce different architectural display settings, such as floor plans, reflected plans, elevations, 3D models, or schematic displays.

3.   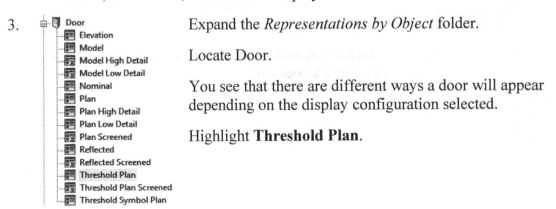   Expand the *Representations by Object* folder.

Locate Door.

You see that there are different ways a door will appear depending on the display configuration selected.

Highlight **Threshold Plan**.

4.

Display Component	Visible
Threshold A	●
Threshold B	●
Threshold A Above Cut Plane	●
Threshold B Above Cut Plane	●
Threshold A Below Cut Plane	●
Threshold B Below Cut Plane	●

Layer/Color/Linetype   Other   Version History

You see that the Threshold visibilities are turned off.

5.

Door
- Elevation
- Model
- Model High Detail
- Model Low Detail
- Nominal
- Plan
- Plan High Detail
- Plan Low Detail

Highlight the **Plan** configuration.

6.

Layer/Color/Linetype   Other   Version History

Display Component	Visible
Panel	●
Frame	●
Stop	●
Swing	●
Direction	●
Panel Above Cut Plane	●
Frame Above Cut Plane	●
Stop Above Cut Plane	●
Swing Above Cut Plane	●
Panel Below Cut Plane	●
Frame Below Cut Plane	●
Stop Below Cut Plane	●
Swing Below Cut Plane	●

You see that some door components are turned on and others are turned off.

7.

Display Theme
Door
- Elevation
- Model
- Model High Detail
- Model Low Detail

Highlight the Model configuration.

8.

Layer/Color/Linetype   Muntins   Other   Versi

Display Component	Visible
Panel	●
Frame	●
Stop	●
Swing	●
Glass	●

You see that some door components are turned on and others are turned off.

9. 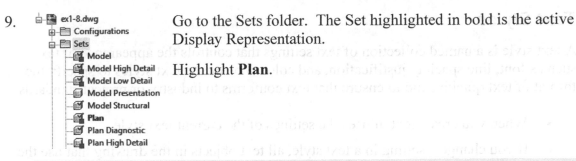   Go to the Sets folder. The Set highlighted in bold is the active Display Representation.

Highlight **Plan**.

10.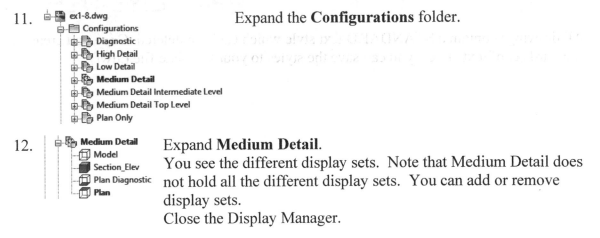

Select the Display Representation Control tab.
Locate **Door** under Objects.
Note that some display representations are checked and others are unchecked.
A check indicates that the door is visible. Which door components are visible is controlled under Representations by Object.

The three different folders – Configurations, Sets, and Representations by Object – are different ways of looking at the same elements. Each folder organizes the elements in a different way. If you change the display setting in one folder, it updates in the other two folders.

11.
```
ex1-8.dwg
  Configurations
    Diagnostic
    High Detail
    Low Detail
    Medium Detail
    Medium Detail Intermediate Level
    Medium Detail Top Level
    Plan Only
```
Expand the **Configurations** folder.

12.
```
Medium Detail
  Model
  Section_Elev
  Plan Diagnostic
  Plan
```
Expand **Medium Detail**.
You see the different display sets. Note that Medium Detail does not hold all the different display sets. You can add or remove display sets.
Close the Display Manager.

13. Close the drawing without saving.

Text Styles

A text style is a named collection of text settings that controls the appearance of text, such as font, line spacing, justification, and color. You create text styles to specify the format of text quickly, and to ensure that text conforms to industry or project standards.

- When you create text, it uses the settings of the current text style.

- If you change a setting in a text style, all text objects in the drawing that use the style update automatically.

- You can change the properties of individual objects to override the text style.

- All text styles in your drawing are listed in the Text Style drop-down.

- Text styles apply to notes, leaders, dimensions, tables, and block attributes.

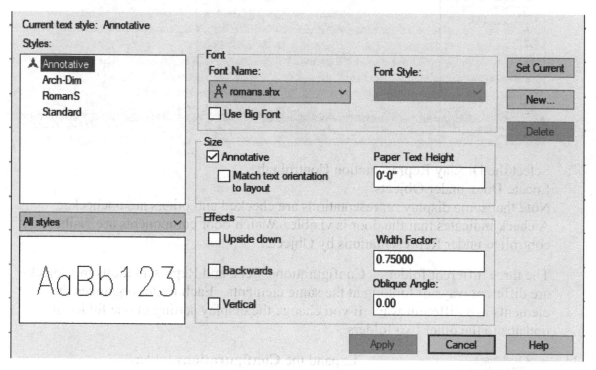

All drawings contain a STANDARD text style which can't be deleted. Once you create a standard set of text styles, you can save the styles to your template file (.dwt).

Exercise 1-10:

Creating a Text Style

Drawing Name: new using QNEW
Estimated Time: 15 minutes

This exercise reinforces the following skills:

- ❑ CUI – Custom User Interface
- ❑ Tab on the ribbon
- ❑ Panel
- ❑ Tab

1. Start a new drawing using the QNEW tool.

2. Select the **Text Style** tool on the Home tab on the ribbon.

    It is located on the Annotation panel under the drop-down.

3. Click the **New** button in the Text Styles dialog box.

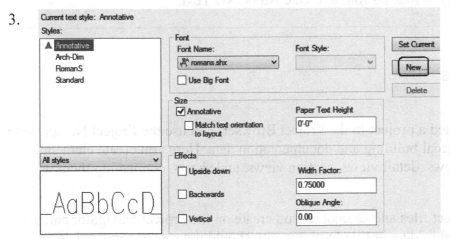

4. In the New Text Style dialog box, enter **TEXT1** for the name of the new text style and Click **OK**.

5.

Current text style: TEXT1

Styles:

A Annotative
Arch-Dim
RomanS
Standard
A TEXT1

All styles

AaBbCcD

Font

Font Name:
Tr Arial

☐ Use Big Font

Font Style:
Regular

Size

☑ Annotative

☐ Match text orientation
to layout

Paper Text Height

0'-0 1/8"

Effects

☐ Upside down

☐ Backwards

☐ Vertical

Width Factor:
1.00000

Oblique Angle:
0.00

Fill in the remaining information in the Text Style dialog box as shown.

Set the Font Name to **Arial**.
Set the Font Style to **Regular**.
Set the Paper Text Height to **1/8″**.
Set the Width Factor to **1.00**.

6. Click the Apply button first and then the Close button to apply the changes to TEXT1 and then close the dialog box.

7. Save as *fonts.dwg*.

Tips Tricks

➤ You can use PURGE to eliminate any unused text styles from your drawing.
➤ You can import text styles from one drawing to another using the Design Center.
➤ You can store text styles in your template for easier access.
➤ You can use any Windows font to create AutoCAD Text.

Project Settings

After you have selected a project in the Project Browser, you use the Project Navigator to create and edit the actual building and documentation data. Here you create elements, constructs, model views, detail views, section views, and sheets, connecting them with one another.

Note: Although project files and categories you create on the Project Navigator palette are shown as files and folders in File Explorer, you should not move, copy, delete, or rename project files from there. Such changes are not updated on the Project Navigator palette, and you could get an inconsistent view of your project data. Any changes you make to the project on the Project Navigator palette are managed and coordinated by the software. Also note that any changes you make on the Project Navigator palette (such as renaming, deleting, or changing the file's properties) cannot be undone using the AutoCAD UNDO command.

Exercise 1-11:
Project Settings

Drawing Name: new using QNEW
Estimated Time: 5 minutes

This exercise reinforces the following skills:

- ❏ Project Navigator
- ❏ Project Path Options

In order for tags on windows, doors, etc. to work properly, you must assign your building model to a project.

1. Start a new drawing by Clicking the + tab.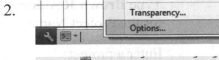

2.  Right click in the command window and select **Options**.

3.  Select the **AEC Project Defaults** tab.

4. Set the AEC Project Location Search Path to the location where you are saving your files.

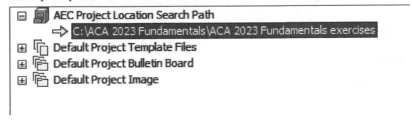

Close the dialog.

5.  Launch the **Project Navigator** from the Quick Access toolbar.

6. Launch the **Project Browser** using the toolbar on the bottom of the Project Navigator palette.

7. 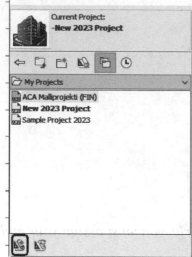 Select the **New Project** tool.

8.

Project Number:

A102

Project Name:

Brown Residence

Project Description:

3 bedroom single family home

☑ Create from template project:

C:\ProgramData\Autodesk\ACA 2020\enu\Template\Commercial Template Project (Imperial)\Cor [...]

[OK]   [Cancel]   [Help]

Type **A102** for the Project Number.

Type **Brown Residence** for the Project Name.

Enter a project description.

Click **OK**.

9. Close the Project Browser.

10. Save as *ex1-11.dwg*.

QUIZ 1

True or False

1. Doors, windows, and walls are inserted into the drawing as objects.

2. The Architectural Toolbars are loaded from the ACA Menu Group.

3. To set the template used by the QNEW tool, use Options.

4. If you start a New Drawing using the 'QNew' tool shown above, the Start-Up dialog will appear unless you assign a template.

5. Mass elements are used to create Mass Models.

6. The Layer Manager is used to organize, sort, and group layers.

7. Layer Standards are set using the Layer Manager.

8. You can assign materials to components in wall styles.

Multiple Choice
Select the best answer.

9. If you start a drawing using one of the standard templates, AutoCAD Architecture will automatically create _____ layouts.

   A. A work layout plus Model Space
   B. Four layouts, plus Model Space
   C. Ten layouts, plus Model Space
   D. Eleven layouts, plus Model Space

10. Options are set through:

    A. Options
    B. Desktop
    C. User Settings
    D. Template

11. The Display Manager controls:

    A. How AEC objects are displayed in the graphics window
    B. The number of viewports
    C. The number of layout tabs
    D. Layers

12. Mass Elements are created using the _____ Tool Palette.

    A. Massing
    B. Concept
    C. General Drafting
    D. Documentation

13. Wall Styles are created by:

    A. Highlighting a Wall on the Wall tab of the Tool Palette, right click and select Wall Styles
    B. Type **WallStyle** on the command line
    C. Go to **Format→Style Manager**
    D. All of the above

14. Visual Styles are used to:

    A. Determine which materials are used for rendering
    B. Controls which layers are visible
    C. Determines the display of edges, lighting, and shading in each viewport
    D. Determines the display of AEC elements in the Work layout tab

15. You can change the way AEC objects are displayed by using:

    A. Edit Object Display
    B. Edit Display Properties
    C. Edit Entity Properties
    D. Edit AEC Properties

16. The Styles Browser can be used for any of the following tasks: (select more than one)

    A. Search for object styles
    B. Import styles into your current drawing
    C. Place an element using a selected style from the browser
    D. Create new styles

ANSWERS:

1) T; 2) T; 3) T; 4) F; 5) T; 6) T; 7) T; 8) T; 9) A; 10) A; 11) A; 12) A; 13) D; 14) C; 15) A; 16) A, B, and C (to create new styles you have to use the Styles Manager)

Lesson 2:
Site Plans

Most architectural projects start with a site plan. The site plan indicates the property lines, the house location, a North symbol, any streets surrounding the property, topographical features, location of sewer, gas, and/or electrical lines (assuming they are below ground – above ground connections are not shown), and topographical features.

When laying out the floor plan for the house, many architects take into consideration the path of sun (to optimize natural light), street access (to locate the driveway), and any noise factors. AutoCAD Architecture includes the ability to perform sun studies on your models.

AutoCAD Architecture allows the user to simulate the natural path of the sun based on the longitude and latitude coordinates of the site, so you can test how natural light will affect various house orientations.

A plot plan must include the following features:

  ❑ Length and bearing of each property line
  ❑ Location, outline, and size of buildings on the site
  ❑ Contour of the land
  ❑ Elevation of property corners and contour lines
  ❑ North symbol
  ❑ Trees, shrubs, streams, and other topological items
  ❑ Streets, sidewalks, driveways, and patios
  ❑ Location of utilities
  ❑ Easements and drainages (if any)
  ❑ Well, septic, sewage line, and underground cables
  ❑ Fences and retaining walls
  ❑ Lot number and/or address of the site
  ❑ Scale of the drawing

The plot plan is drawn using information provided by the county/city and/or a licensed surveyor.

When drafting with pencil and vellum, the drafter will draw to a scale, such as ⅛″ = 1′, but with AutoCAD Architecture, you draw full-size and then set up your layout to the proper scale. This ensures that all items you draw will fit together properly. This also is important if you collaborate with outside firms. For example, if you want to send your files to a firm that does HVAC, you want their equipment to fit properly in your design. If everyone drafts 1:1 or full-size, then you can ensure there are no interferences or conflicts.

Linetypes

Architectural drafting requires a variety of custom linetypes in order to display specific features. The standard linetypes provided with AutoCAD Architecture are insufficient from an architectural point of view. You may find custom linetypes on Autodesk's website, www.cadalog.com, or any number of websites on the Internet. However, the ability to create linetypes as needed is an excellent skill for an architectural drafter.

It's a good idea to store any custom linetypes in a separate file. The standard file for storing linetypes is acad.lin. If you store any custom linetypes in acad.lin file, you will lose them the next time you upgrade your software.

Exercise 2-1:

Creating Custom Line Types

Drawing Name: New
Estimated Time: 15 minutes

This exercise reinforces the following skills:

- ❑ Creation of linetypes
- ❑ Customization

A video of this lesson is available on my MossDesigns channel on Youtube at https://www.youtube.com/watch?v=gSCyFJm6pfY

Start a new drawing using **QNEW**.

1. ———————————   ————   ————   ———————————

    Draw a property line.
    This property line was created as follows:
        Set ORTHO ON.
        Draw a horizontal line 100 units long.

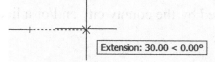

Extension: 30.00 < 0.00°

    Use SNAP FROM to start the next line @30,0 distance from the previous line.
    You can also simply use object tracking to locate the next line.
    The short line is 60 units long.
    Use SNAP FROM to start the next line @30,0 distance from the previous line.
    The second short line is 60 units long.
    Use SNAP FROM to start the next line @30,0 distance from the previous line.
    Draw a second horizontal line 100 units long.

2. 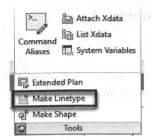 Activate the Express Tools tab on the ribbon.

Select the **Tools** drop-down.

Go to **Make Linetype**.

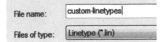

3. Browse to the folder where you are storing your work.

File name: custom-linetypes

Files of type: Linetype (*.lin)

Create a file name called *custom-linetypes.lin* and click **Save**.

Place all your custom linetypes in a single drawing file and then use the AutoCAD Design Center to help you locate and load the desired linetype.

4. When prompted for the linetype name, type property-line.
   When prompted for the linetype description, type property-line.
   Specify starting point for line definition; select the far left point of the line type.
   Specify ending point for line definition; select the far right point of the line type.
   When prompted to select the objects, start at the long line on the left and pick the line segments in order.

5. Type **linetype** to launch the Linetype Manager.

6.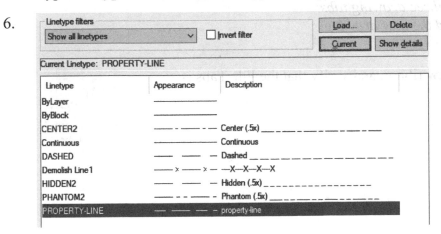

The linetype you created is listed.

Highlight the PROPERTY-LINE linetype and select CURRENT.

Click **OK**.

7. Draw some lines to see if they look OK. Make sure you make them long enough to see the dashes. If you don't see the dashes, adjust the linetype scale using Properties. Set the scale to 0.0125 or smaller.

8. 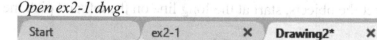 custom-linetypes   Locate the *custom-linetypes.lin* file you created.

    Open it using NotePad.

9.

custom linetypes - Notepad
File Edit Format View Help
*PROPERTY-LINE,property line
A,200,-60,120,-60,200,-60,120,-60,120|

You see a description of the property-line.

10. Save as *ex2-1.dwg*.

Exercise 2-2:
Creating New Layers

Drawing Name: New
Estimated Time: 20 minutes

This exercise reinforces the following skills:

- ❏ Design Center
- ❏ Layer Manager
- ❏ Creating New Layers
- ❏ Loading Linetypes

1. 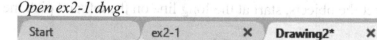 Start a new drawing using QNEW.

    Open ex2-1.dwg.

Start	ex2-1	✕	**Drawing2***	✕

You should see three drawing tabs:
Start, ex2-1, and the new drawing

Activate the **View** ribbon and verify that the File Tabs is shaded.

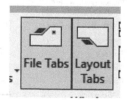

2.   Launch the Design Center.
     The Design Center is on the Insert ribbon in
     the Content section. It is also located in the
     drop-down under the Content Browser. You
     can also launch using Ctl+2.

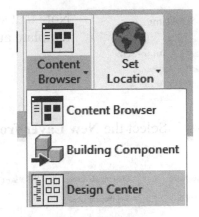

3.

Select the **Open Drawings** tab.

Browse to the *ex2-1.dwg* file.

Open the Linetypes folder.

Scroll to the **Property-line** linetype
definition.

4.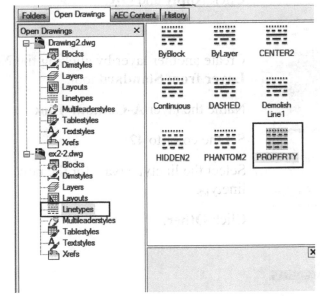

Highlight the PROPERTY-LINE
linetype.

Drag and drop into your active
window or simply drag it onto the
open drawing listed.

*This copies the property-line
linetype into the drawing.*

5.   Close the Design Center.

6.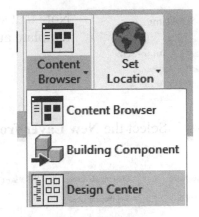

Activate the **Layer Properties Manager** on the
Home ribbon.

7. 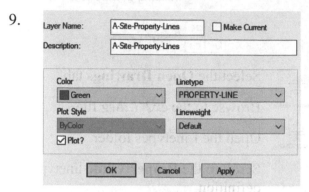 Note that several layers have already been created. The template automatically sets your Layer Standards to AIA.

8. Select the **New Layer from Standard** tool.

9. 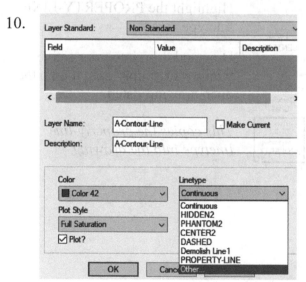 Under Layer Standard, select **Non Standard**.

Name the layer **A-Site-Property-Lines**.

Set the Color to **Green**.

Select the **Property-Line** linetype.

Set the Lineweight to **Default**.

Enable **Plot**.
Click **Apply** and **OK**.

10.

Create another layer by clicking the **New Layer from Standard** tool.

Name the layer **A-Contour-Line**.

Set the color to **42**.

Select the linetype pane for the new linetype.

Click **Other**.

11. 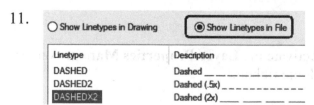 Enable **Show Linetypes in File**.

Highlight **DASHEDX2** and click **OK** to assign it to the selected layer name.

12.

| Layer Name: | A-Contour-Line | ☐ Make Current |
| Description: | A-Contour-Line | |

Color
■ Color 42 ∨

Linetype
DASHEDX2 ∨

Plot Style
ByColor ∨

Lineweight
Default ∨

☑ Plot?

OK   Cancel   Apply

Note that the linetype was assigned properly.

Set the Lineweight to **Default**.

Enable **Plot**.

Click **OK**.

13.

S.. Name
⟋ 0
⟋ A-Contour-Line
✓ A-Site-Property-Lines
⟋ Defpoints

Set the A-Site-Property-Line layer current.

Highlight the layer name then select the green check mark to set current.

You can also double left click to set the layer current.

14. Right click on the header bar and select **Maximize all columns** to see all the properties for each layer.

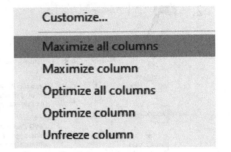

Customize...
Maximize all columns
Maximize column
Optimize all columns
Optimize column
Unfreeze column

15. Use the X in the upper corner to close the Layer Manager.

16. Save as *ex2-2.dwg*.

Exercise 2-3:
Creating a Site Plan

Drawing Name: site_plan.dwg
Estimated Time: 30 minutes

This exercise reinforces the following skills:

- ❑ Use of ribbons
- ❑ Documentation Tools
- ❑ Elevation Marks
- ❑ Surveyors' Angles

1.   Open *site_plan.dwg*.

2. 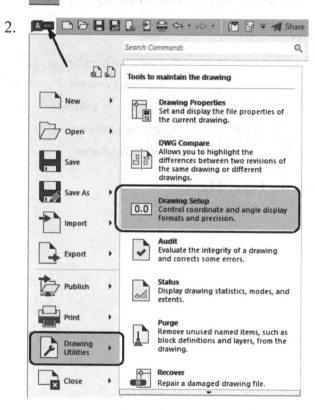 Go to **Drawing Utilities→Drawing Setup**.

This is located on the Application Menu panel.

3.

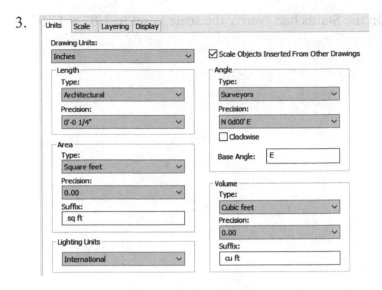

Select the **Units** tab.
Set the Drawing Units to
Inches.
Set up the Length Type to
Architectural.
Set the Precision to ¼″.
Set the Angle Type to
Surveyors.
Set the Precision to **N 0d00′ E**.

4.

Select the **Scale** tab.

Set the Scale to **1/8" = 1'-0"**.

Set the Annotation Plot Size to
¼".

Click **OK**.

5.

You have changed the units for this drawing, which will be used on all new items. What do you want to do?

→ Rescale modelspace and paperspace objects
   Make all match the new units

→ Rescale only modelspace objects
   Make them match the new units

→ Don't rescale any existing objects

When this dialog appears, select the first option. The drawing doesn't have any pre-existing elements, but this option ensures that everything is scaled correctly.

6.

On the Status bar: verify the scale is set to **1/8" = 1'-0"**.

You can set your units as desired and then enable **Save As Default** ☐ Save As Default .
Then, all future drawings will use your unit settings.

---

7. Draw the property line shown on the **A-Site-Property-Lines** layer.

Select the Line tool.

8.

Specify first point: 1',5'
Specify next point: @186' 11" < s80de
Specify next point: @75'<n30de
Specify next point: @125'11"<n
Specify next point: @250'<180

This command sequence uses Surveyor's Units.

You can use the TAB key to advance to the surveyor unit box.

```
Specify first point: 12,60

Specify next point or [Undo]: @2243<s80de

Specify next point or [Undo]: @900<n30de

Specify next point or [Close/Undo]: @1511<n

Specify next point or [Close/Undo]: @3000<180

Specify next point or [Close/Undo]:
```

Metric data entry using centimeters:

If you double click the mouse wheel, the display will zoom to fit so you can see your figure.

START

Turn ORTHO off before creating the arc.

9. 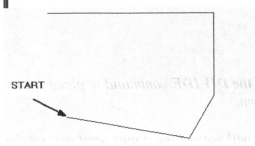 Draw an arc using **Start, End, Direction** to close the figure.

10. 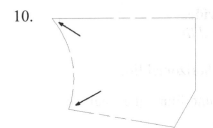 Select the bottom point as the Start point.
Right click and select the End option.
Select the top point at the End point.
Click inside the figure to set the direction of the arc.

11. Set the **A-Contour-Line layer** current.

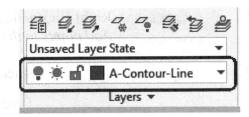

12. PTYPE Type **PTYPE** to access the Point Style dialog.

13. Select the Point Style indicated.

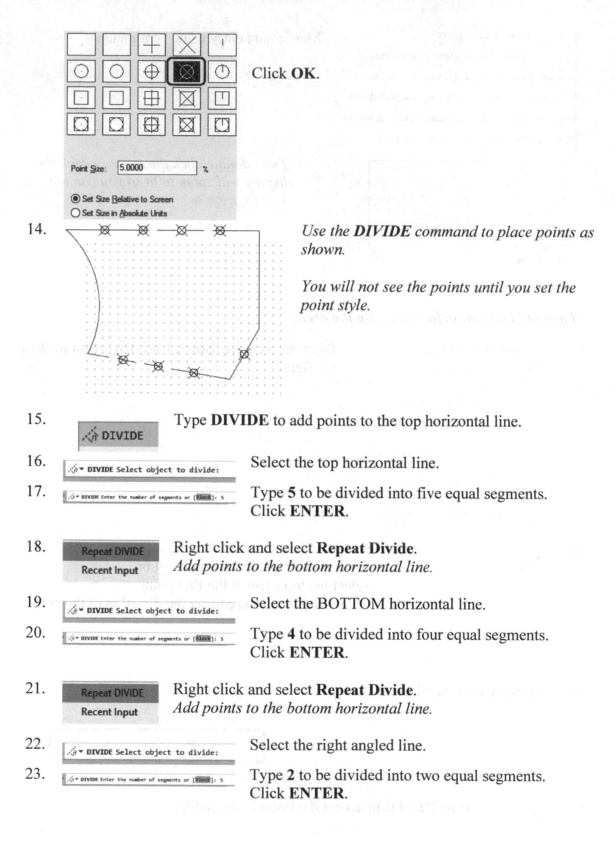

Click **OK**.

14.   *Use the **DIVIDE** command to place points as shown.*

You will not see the points until you set the point style.

15.   Type **DIVIDE** to add points to the top horizontal line.

16.   Select the top horizontal line.

17.   Type **5** to be divided into five equal segments. Click **ENTER**.

18.   Right click and select **Repeat Divide**.
Add points to the bottom horizontal line.

19.   Select the BOTTOM horizontal line.

20.   Type **4** to be divided into four equal segments. Click **ENTER**.

21.   Right click and select **Repeat Divide**.
Add points to the bottom horizontal line.

22.   Select the right angled line.

23.   Type **2** to be divided into two equal segments. Click **ENTER**.

24. 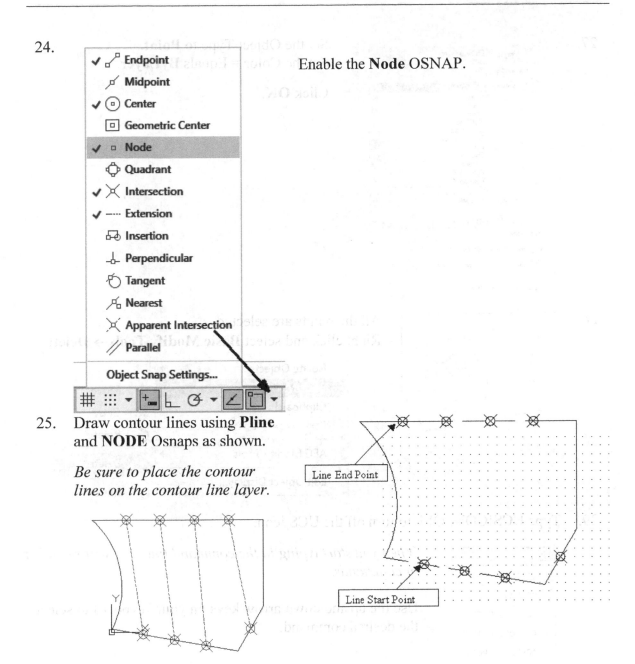 Enable the **Node** OSNAP.

25. Draw contour lines using **Pline** and **NODE** Osnaps as shown.

    Be sure to place the contour lines on the contour line layer.

26. Type **qselect** on the *Command* line.

27.

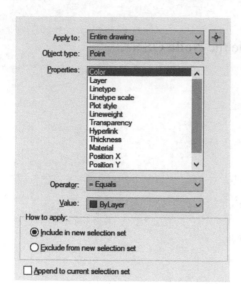

Set the Object Type to **Point**.
Set the Color = Equals **ByLayer**.

Click **OK**.

28.

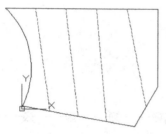

All the points are selected.
Right click and select **Basic Modify Tools→ Delete**.

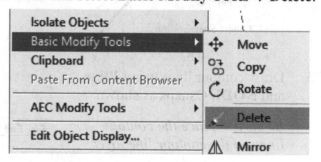

29. Type UCSICON, OFF to turn off the UCS icon.

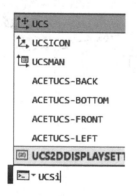

When you start typing in the command line, you will see a list of commands.

Use the up and down arrow keys on your keyboard to select the desired command.

30.

Activate the **Design Tools** palette located under the Tools drop-down on the Home ribbon.

31. 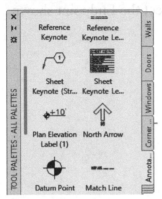 Right click on the gray bar on the palette to activate the short-cut menu.

Left click on **All Palettes** to enable the other palettes.

32. Activate the Annotation tab on the palette.

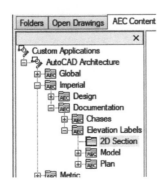 Locate the **Plan Elevation Label**, but it's not the type we need for a site plan.

Instead we'll have to add an elevation label from the Design Center to the Tool Palette.

33. 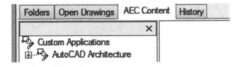 Type **Ctl+2** to launch the Design Center.
Or:
Activate the **Insert** ribbon.
Select the **Design Center** tool under Content.
Or:
Type **DC**.

34. Select the **AEC Content** tab.

35. Browse to
Imperial/Documentation/Elevation Labels/ 2D Section.

[Metric/Documentation/E levation Labels/2D Section].

36. 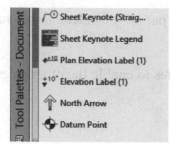 Locate the Elevation Label (1) file.

Highlight the elevation label.

Drag and drop onto the Tool Palette.

Close the Design Center.

37. 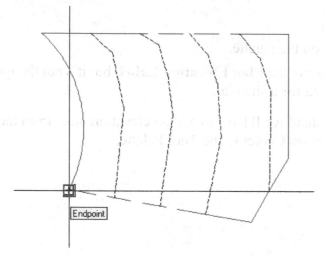 Select the Elevation Label on the Tool Palette.

Select the Endpoint shown.

38. Set the Elevation to **0″ [0.00]**.
Set the Prefix to **EL**.
Click **OK**.

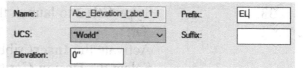

39.   Zoom in to see the elevation label.

 You can use the **Zoom Object** tool to zoom in quickly.

This is located on the View tab.

The Elevation Label is automatically placed on the G-Anno-Dims layer.

To verify this, just hover over the label and a small dialog will appear displaying information about the label.

40. EL ±0.00"   Place an elevation label on the upper left vertex as shown.

41.   Place labels as shown.

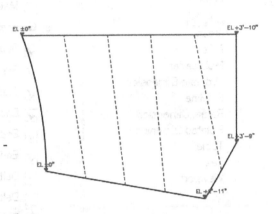

42.

Scale	
X	36.00000
Y	36.00000
Z	36.00000

To change the size of the labels, select and set the Scale to **36.00** on the Properties dialog.

43.

Edit Object Display...
Elevation Label Modify...
Edit Multi-View Block Definition...

To modify the value of the elevation label, select the label.

Right click and select **Elevation Label Modify**.

44.

Geometry	
Start X	1'-0"
Start Y	5'-0"
Start Z	0"
End X	185'-0 15/16"
End Y	-27'-5 1/2"
End Z	2'-11"
Delta X	184'-0 15/16"
Delta Y	-32'-5 1/2"

To change the toposurface to a 3D surface, select each line and modify the Z values to match the elevations.

45. *A quick way to modify properties is to use Quick Properties.*

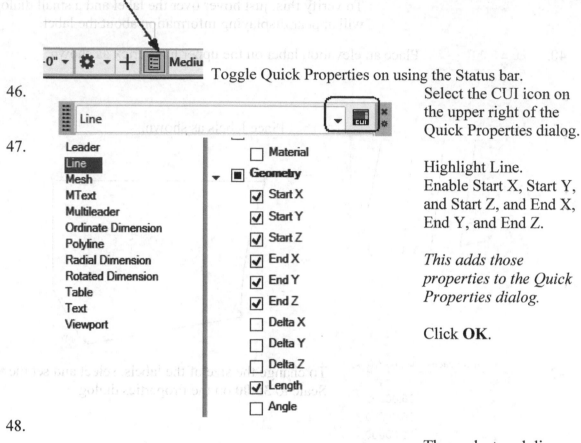

Toggle Quick Properties on using the Status bar.

46.   Select the CUI icon on the upper right of the Quick Properties dialog.

47.   Highlight Line.
Enable Start X, Start Y, and Start Z, and End X, End Y, and End Z.

This adds those properties to the Quick Properties dialog.

Click **OK**.

48.

Then select each line and change the start and end values of Z to match the elevation markers.

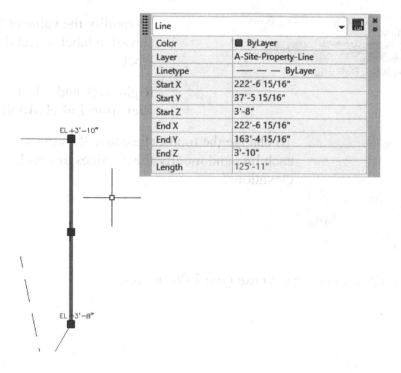

49.  Use the ViewCube to inspect your toposurface and then return to a top view.

Adjust points as needed to create an enclosed area. Use ID to check the point values of each end point.

You can also type PLAN to return to a 2D view plan view.

50. Open the *fonts.dwg*. This is the file you created in a previous exercise.

51. Type **Ctl+2** to launch the Design Center.

52. Select the Open Drawings tab.

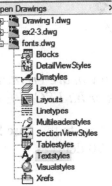

Browse to the Textstyles folder under the fonts.dwg.

53.  Drag and drop the **Text1** text style into the ex2-3 drawing.

Close the Design Center.

54.  Set **G-Anno-Dims** as the current Layer.

55. 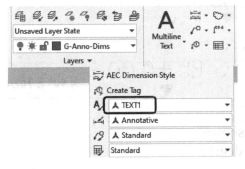 Set the Current Text Style to **TEXT1.**

56. 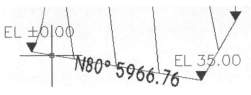 Select the **Single Line** text tool from the Home ribbon.

57. EL ±0|00

N80° 5966.76   EL 35.00

N80° 186' 11"

Right click and select **Justify**.
Set the Justification to **Center**.
Select the midpoint of the line as the insertion point.
Set the rotation angle to **–10** degrees.
Use %%d to create the degree symbol.

58.  To create the S60… note, use the **TEXT** command.

Use a rotation angle of 60 degrees.

When you are creating the text, it will preview horizontally. However, once you left click or hit ESCAPE to place it, it will automatically rotate to the correct position.

59.  Create the **Due South 125′ 11″ [Due South 3840.48]** with a rotation angle of 90 degrees.

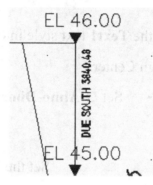

Add the text shown on the top horizontal property line.

N 90° 250′ E

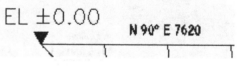

60. 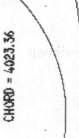 Add the Chord note shown.
Rotation angle is 90 degrees.

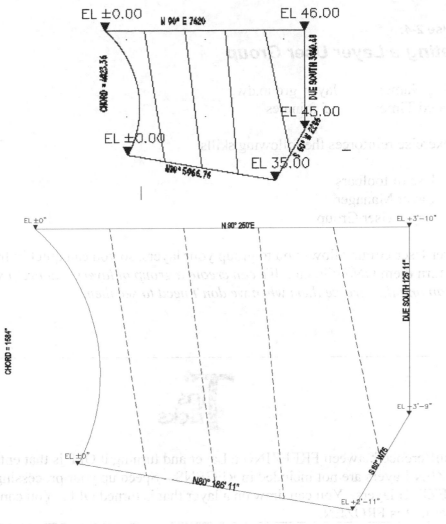

61. Adjust the scale to **1/32″ = 1′-0″** to see the
     property line labels.

| 1/64" = 1'-0" |
| 1/32" = 1'-0" |
| 1/16" = 1'-0" |
| 3/32" = 1'-0" |
| 1/8" = 1'-0" |
| 3/16" = 1'-0" |
| 1/4" = 1'-0" |
| 3/8" = 1'-0" |
| 1/2" = 1'-0" |
| 3/4" = 1'-0" |
| 1" = 1'-0" |
| 1 1/2" = 1'-0" |
| 3" = 1'-0" |
| 6" = 1'-0" |
| 1'-0" = 1'-0" |
| Custom... |
| Hide Xref scales |

62. Save the file as *ex2-3.dwg*.

Exercise 2-4:

Creating a Layer User Group

Drawing Name: layer_group.dwg
Estimated Time: 15 minutes

This exercise reinforces the following skills:

- ❑ Use of toolbars
- ❑ Layer Manager
- ❑ Layer User Group

A Layer User Group allows you to group your layers, so you can quickly freeze/thaw them, turn them ON/OFF, etc. *We can create a group of layers that are just used for the site plan and then freeze them when we don't need to see them.*

The difference between FREEZING a Layer and turning it OFF is that entities on FROZEN Layers are not included in REGENS. Speed up your processing time by FREEZING layers. You can draw on a layer that is turned OFF. You cannot draw on a layer that is FROZEN.

1.   📂   Open *layer_group.dwg*.

2.   📑   Select the **Layer Properties Manager** tool from the Home tab on the ribbon.

3.   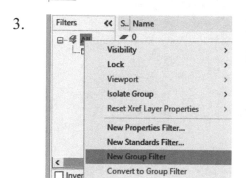 Highlight the word **All** in the Filters pane.

      Right click and select the **New Group Filter**.

4. 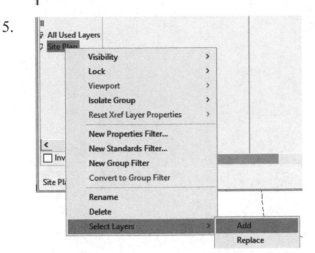 Name the group **Site Plan**.

5. Highlight the Site Plan group.
Right click and select **Select Layers →**
Add.

6. Type **ALL** on the command line.
This selects all the items in the drawing.
Click **ENTER** to finish the selection.
The layers are now listed in the Layer Manager under the Site Plan group.

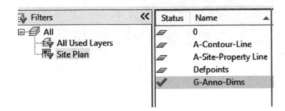

Close the Layer Manager.

7. We can use this group to quickly freeze, turn off, etc. the layers in the Site Plan group.

Save the drawing as *ex2-4.dwg* and close.

Notes:

QUIZ 2

True or False

1. You must load a custom linetype before you can assign it to a layer.
2. The DIVIDE command divides a line or arc segment into smaller elements.
3. The Design Tools palette is used to access blocks and commands.
4. If you design a wall tool on the tool palette for a specific style wall, then whenever you add a wall using that design tool it will place a wall using that wall style.
5. Layer groups control which layers are displayed in the layer drop-down list.

Multiple Choice
Select the best answer.

6. If points are not visible in a view:

    A. Turn on the layer the points are placed on
    B. Bring up the Point Style dialog and set the Size Relative to Screen
    C. Change the Point Style used
    D. All of the above

7. The command to bring up the point style dialog is:

    A. PSTYLE
    B. PMODE
    C. PP
    D. STYLE

8. An easy way to change the units used in a drawing is to:

    A. Change the units on the status bar.
    B. Type UNITS
    C. Use the Drawing Setup dialog.
    D. B or C

9. To turn off the display of the USCICON:

    A. Type USCICON, OFF
    B. Type UCS, OFF
    C. Type UCSICON, 0
    D. Type UCS, 0

ANSWERS:
1) T; 2) F; 3) T; 4) T; 5) T; 6) D; 7) A; 8) D; 9) A

Lesson 3:
Floor Plans

AutoCAD Architecture comes with 3D content that you use to create your building model and to annotate your views. In ACA 2023, you may have difficulty locating and loading the various content, so this exercise is to help you set up ACA so you can move forward with your design.

The Content Browser lets you store, share, and exchange AutoCAD Architecture content, tools, and tool palettes. The Content Browser runs independently of the software, allowing you to exchange tools and tool palettes with other Autodesk applications.

The Content Browser is a library of tool catalogs containing tools, tool palettes, and tool packages. You can publish catalogs so that multiple users have access to standard tools for projects.

ACA comes with several tool catalogs. When you install ACA, you enable which catalogs you want installed with the software. By default, Imperial, Metric, and Global are enabled. The content is located in the path: C:\ProgramData\Autodesk\ACA 2023\enu\Tool Catalogs.

The floor plan is central to any architectural drawing. In the first exercise, we convert an AutoCAD 2D floor plan to 3D. In the remaining exercises, we work in 3D.

A floor plan is a scaled diagram of a room or building viewed from above. The floor plan may depict an entire building, one floor of a building, or a single room. It may also include measurements, furniture, appliances, or anything else necessary to the purpose of the plan.

Floor plans are useful to help design furniture layout, wiring systems, and much more. They're also a valuable tool for real estate agents and leasing companies in helping sell or rent out a space.

Exercise 3-1:

Going from a 2D to 3D Floor plan

Drawing Name: New
Estimated Time: 45 minutes

This exercise reinforces the following skills:

- ☐ Create Walls
- ☐ Wall Properties
- ☐ Wall Styles
- ☐ Style Manager
- ☐ Insert an AutoCAD drawing
- ☐ Trim, Fillet, Extend Walls

1. 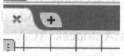 Start a new drawing using QNEW or select the + tab.

2. Type **UNITS.**

    Set the Units to **Inches.**

    Set the Type to **Architectural.**

    Set the Precision to ¼".

    Click **OK.**

3. You have changed the units for this drawing, which will be used on all new items. What do you want to do?

    → Rescale modelspace and paperspace objects
    Make all match the new units

    → Rescale only modelspace objects
    Make them match the new units

    → Don't rescale any existing objects

    Select the first option: **Rescale modelspace and paperspace objects.**

4.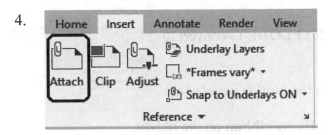

Activate the **Insert** tab on the ribbon.

Select **Attach**.

5.

| File name: | autocad_floor_plan |
| Files of type: | Drawing (*.dwg) |

Locate the *autocad_floor_plan.dwg* file in the exercises. *Set your Files of type to Drawing (*.dwg) to locate the file.* Click **Open**.

6.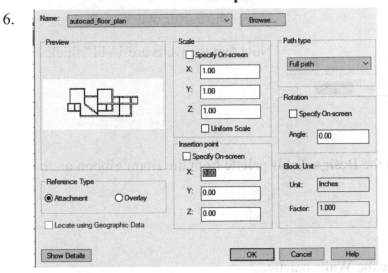

Uncheck Insertion Point.
Uncheck Scale.
Uncheck Rotation.
This sets everything to the default values.

Click **OK**.

7.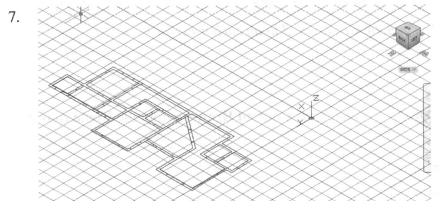

Use the ViewCube to switch to a 3D view.

Note that the AutoCAD file is 2D only.

Return to a top view.

8.

Select the attached xref.
Right click and select **Bind→Insert**.
This converts the xref to an inserted block.

9. Select the block reference and type **EXPLODE** to convert to lines.

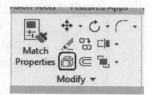

You can also use the Explode tool on the Modify panel of the Home ribbon on the ribbon.

10. 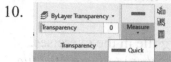 Activate the **Home ribbon** on the ribbon.

Select the **Quick** measure tool from the Utilities panel.

11. Hover the cursor over a wall.

Note that the walls are 1'-11" thick.

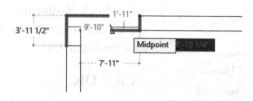

12.  Launch the **Design Tools** palette from the Home ribbon on the ribbon.

13. 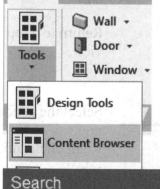 Activate the **Walls** palette.

This palette was created in an earlier exercise.

14. Launch the **Content Browser** from the Home ribbon.

15. In the Search field, type **Stud wall**.

Click **Go**.

16.

Click **Next** until you locate the **Stud-4 Rigid-1.5 Air-1 Brick-4** wall style.

17.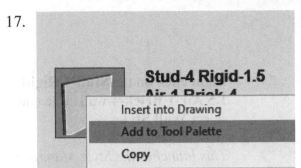

Right click on the desired wall style.

Select **Add to Tool Palette**.

18. Close the Content Browser.

19.

Place the wall style in the area used by exterior wall styles.

20.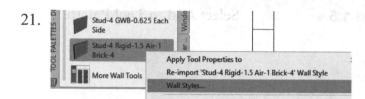

Right click on the **Stud-4 Rigid-1.5 Air-1 Brick-4** wall style and select **Import Stud-4 Rigid-1.5 Air-1 Brick-4 wall style.**

This makes the wall style available in the active drawing.

21.

Right click on the **Stud-4 Rigid-1.5 Air-1 Brick-4** wall style and select **Wall Styles**.

This launches the Style Manager.

22.

Note that the only wall styles available are Standard, the style you saved in the template, and the style you just imported.

Highlight the **Stud-4 Rigid-1.5 Air-1 Brick-4** wall style.

23.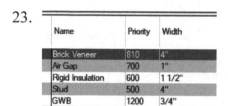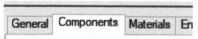

Name	Priority	Width
Brick Veneer	810	4"
Air Gap	700	1"
Rigid Insulation	600	1 1/2"
Stud	500	4"
GWB	1200	3/4"

Activate the **Components** tab.

The components tab lists the materials used in the wall construction.

Total Width: 11 1/4"

Note the components listed in the Style Manager for the wall style. The total wall thickness is 11-1/4".

The total width value is located in the upper right of the dialog. We need a wall style that is 1′–11″. We need to add material to the wall style.

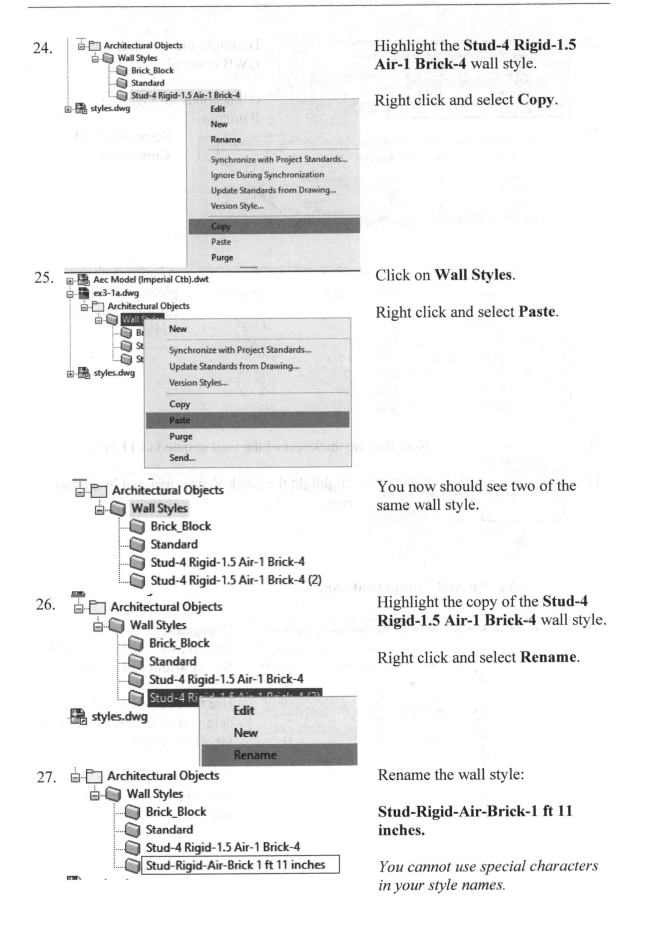

24. Highlight the **Stud-4 Rigid-1.5 Air-1 Brick-4** wall style.

Right click and select **Copy**.

25. Click on **Wall Styles**.

Right click and select **Paste**.

You now should see two of the same wall style.

26. Highlight the copy of the **Stud-4 Rigid-1.5 Air-1 Brick-4** wall style.

Right click and select **Rename**.

27. Rename the wall style:

Stud-Rigid-Air-Brick-1 ft 11 inches.

You cannot use special characters in your style names.

28. 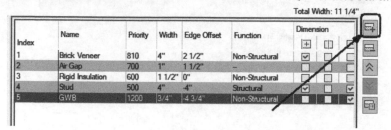 Highlight the row that lists the GWB material.

GWB stands for Gypsum Wallboard.

29. Select the **Add Component** tool.

30.

Index	Name	Priority	Width	Edge Offset
1	Brick Veneer	810	4"	2 1/2"
2	Air Gap	700	1"	1 1/2"
3	Rigid Insulation	600	1 1/2"	0"
4	Stud	500	4"	-4"
5	GWB	1200	3/4"	-3 1/2"
6	GWB	1200	3/4"	-4"

Another 3/4" piece of GWB (gypsum board) is added.

31. Note that the thickness of the wall updated to **11 ¼".**

Total Width: 11 1/4"

32. 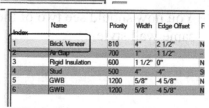 Highlight the Brick Veneer material in the top row.

33. Select the **Add Component** tool.

34. 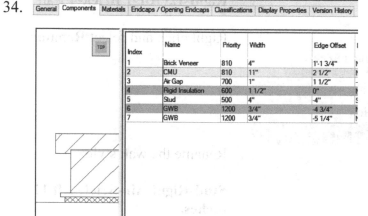 Change the name of the second row material to **CMU**. Set the width to 11" thick.

Use the Preview window to help you adjust the edge offset for all the components.

To change the values, just place the cursor in that cell and start typing.

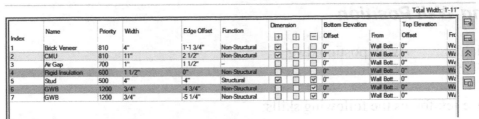

Total Width: 1'-11"

Index	Name	Priority	Width	Edge Offset	Function	Dimension			Bottom Elevation		Top Elevation	
						⊞	⊡	⊟	Offset	From	Offset	Fr
1	Brick Veneer	810	4"	1'-1 3/4"	Non-Structural	☑	☐	☐	0"	Wall Bott...	0"	Wa
2	CMU	810	11"	2 1/2"	Non-Structural	☑	☐	☐	0"	Wall Bott...	0"	Wa
3	Air Gap	700	1"	1 1/2"	–	☐	☐	☐	0"	Wall Bott...	0"	Wa
4	Rigid Insulation	600	1 1/2"	0"	Non-Structural	☐	☐	☐	0"	Wall Bott...	0"	Wa
5	Stud	500	4"	-4"	Structural	☑	☐	☑	0"	Wall Bott...	0"	Wa
6	GWB	1200	3/4"	-4 3/4"	Non-Structural	☐	☐	☑	0"	Wall Bott...	0"	Wa
7	GWB	1200	3/4"	-5 1/4"	Non-Structural	☐	☐	☑	0"	Wall Bott...	0"	Wa

35. Verify that your layers are set as shown.
    Verify that the total width is 1' 11".

36.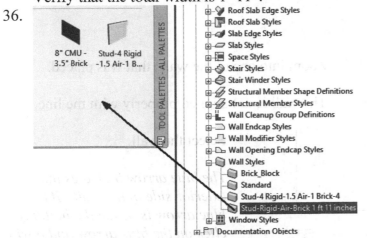

Drag and drop the new wall style onto the Walls tool palette.

37. Click **OK** to close the Styles Manager dialog.

38. If you are prompted to save the drawing file, save as *ex3-1.dwg*.

39.

Right click on the **Stud-Rigid-Air-Brick-1 ft 11 inches** wall style and select **Apply Tool Properties to → Linework**.

40.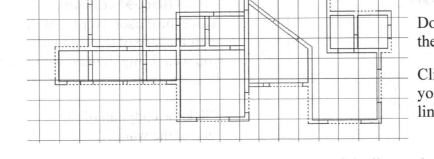

Select the outside segments of the walls.

Do not select any of the interior walls.

Click ENTER when you are done selecting lines.

41. You will be prompted if you want to erase any of the linework. Enter **NO**.

The walls placed are misaligned. We will fix that in the next exercise.

Exercise 3-2:

Adjusting Wall Position

Drawing Name: wall_postion.dwg
Estimated Time: 15 minutes

This exercise reinforces the following skills:

- ❏ Reposition Walls using Offset From
- ❏ Wall Properties

1. Zoom into one of the walls that was placed.

   The wall is not aligned properly with the line.

2. Select the wall.

   The blue arrow indicates the exterior side of the wall. If the blue arrow is inside the building, click on the blue arrow and it will flip the orientation of the wall.

3. *When the wall is selected, the ribbon will change to a context sensitive ribbon for the selected element.*

   On the Walls ribbon/Modify panel:

   Click **Offset→Set From**.

4. Move the cursor over the wall until the face of the component you want is highlighted with a red line, and click once.

   Click the endpoint of the outside face of the wall.

5.

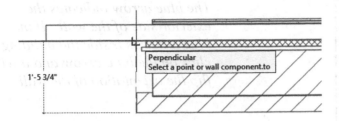

1'-5 3/4"

Perpendicular
Select a point or wall component.to

Click a point on the line perpendicular to the selected line.

6.

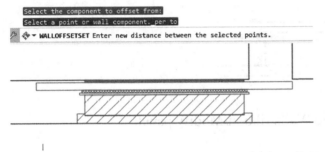

0"

Enter new distance betwee

Enter **0"** when prompted for the distance between the selected points.

Click ESC to release the selection.

Select the component to offset from:
Select a point or wall component. per to
WALLOFFSETSET Enter new distance between the selected points.

Note that it is the correct width.

7.

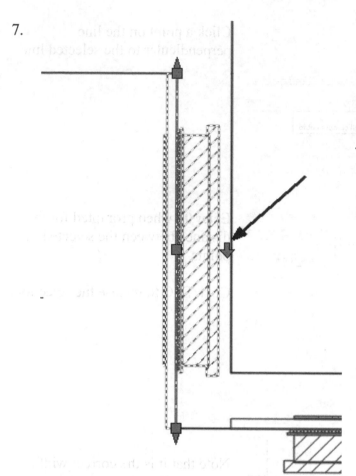

Select another wall.

The blue arrow indicates the exterior side of the wall. If the blue arrow is inside the building, click on the blue arrow and it will flip the orientation of the wall.

Click on the blue arrow to flip the orientation of the wall.

8.

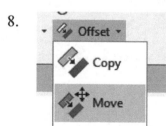

On the Walls ribbon/Modify panel:

Click **Offset→Move**.

9.

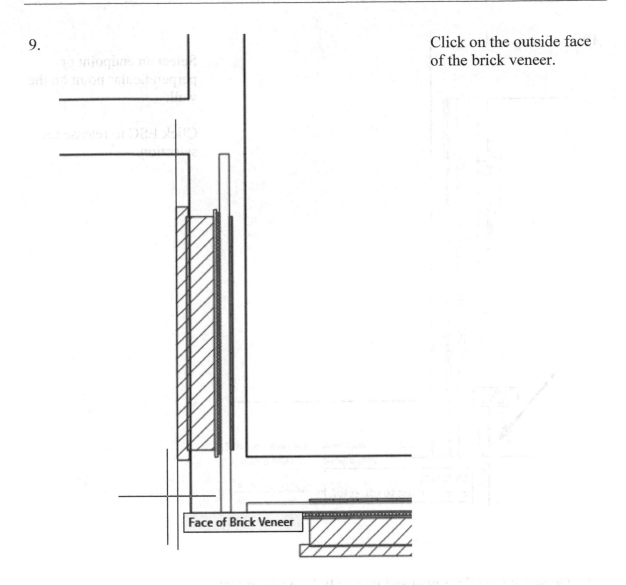

Face of Brick Veneer

Click on the outside face of the brick veneer.

10.

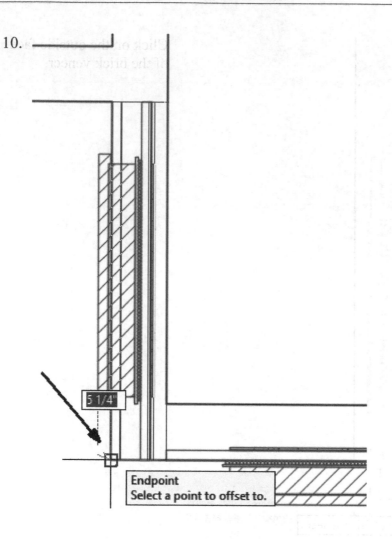

5 1/4"

Endpoint
Select a point to offset to.

Select an endpoint or perpendicular point on the wall.

Click ESC to release the selection.

11. Go around the floor plan and move all the walls so that they overlay the floor plan correctly using the Offset tool.

12.

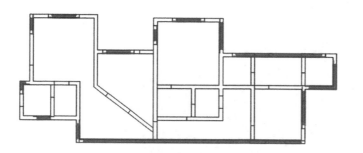

The floor plan should look similar to this. Yours may look slightly different depending on which lines you selected when you placed the walls.

Save as ex3-2.dwg.

Exercise 3-3:
Joining Walls

Drawing Name: wall_fillet.dwg
Estimated Time: 15 minutes

This exercise reinforces the following skills:

- Connecting Walls using Fillet
- Wall Properties

1.

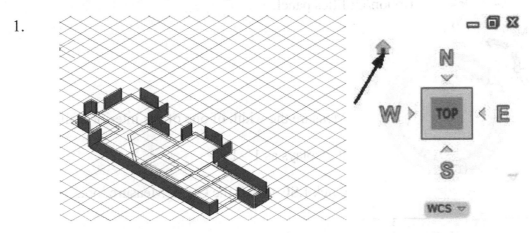

Switch to a 3D view.

You should see 3D walls where you selected lines.

2.

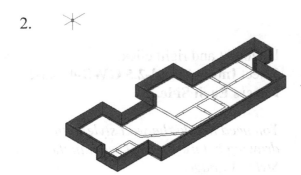

To join the walls together, use FILLET with an R value of 0.
Type FILLET, then select the two walls to be joined to form a corner.

3.

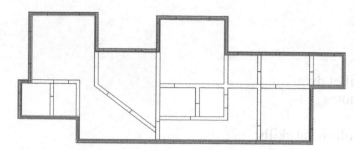

In the plan view, the exterior walls should form a closed figure.

4.

Click on the **Quick Measure** tool located on the Home ribbon/Utilities panel.

5.

Measure the width of one of the interior walls.

It displays as 1' -11".

Click ESC to exit the Quick Measure tool.

6.

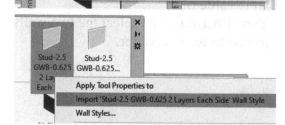

Locate the **Stud-2.5 GWB-0.625-2 Layers Each Side** Wall Style on the Walls tool palette.

7.

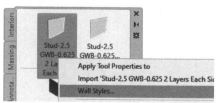

Highlight and right click.
Select **Import Stud-2.5 GWB-0.625-2 Layers Each Side.**

You need to add the wall style to the drawing before you will see it in the Styles Manager.

8.

Right click and select Wall Styles.
This will launch the Styles Manager.

9. Highlight the **Stud-2.5 GWB-0.625-2 Layers Each Side** Wall Style in the Style Manager list.

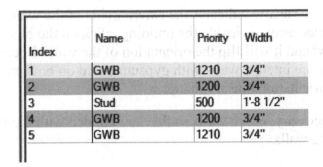

Wall Styles
- Brick_Block
- Standard
- Stud-2.5 GWB-0.625 2 Layers Each Side
- Stud-4 Rigid-1.5 Air-1 Brick-4
- Stud-Rigid-Air-Brick 1 ft 11 inches

Select the **Components** tab.

The total width for this wall style is 5"

10. Change the Stud width to **1' 8.5"**.

Index	Name	Priority	Width
1	GWB	1210	3/4"
2	GWB	1200	3/4"
3	Stud	500	1'-8 1/2"
4	GWB	1200	3/4"
5	GWB	1210	3/4"

11. Adjust the positions of the components so that the wall looks proper.

Index	Name	Priority	Width	Edge Offset	F
1	GWB	1210	3/4"	1'-9"	N
2	GWB	1200	3/4"	1'-8 1/2"	St
3	Stud	500	1'-8 1/2"	0"	St
4	GWB	1200	3/4"	-3/4"	N
5	GWB	1210	3/4"	-1 1/4"	N

12. Verify that the total width is 1'-11".

Total Width: 1'-11"

13. Click **OK** to close the Style Manager.

14. Select the **Stud-4 GWB-0.625-2 Layers Each Side** wall style.

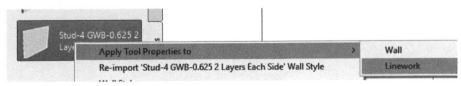

Right click and select **Apply Tool Properties to → Linework**.

15.

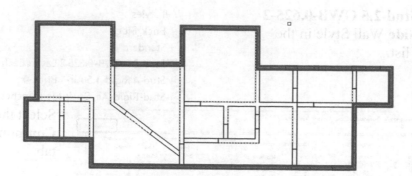

Select the inside segments of the walls.

Do not select any of the exterior walls.

Click ENTER when you are done selecting lines.

16. You will be prompted if you want to erase any of the line work. Enter **NO**.

17.

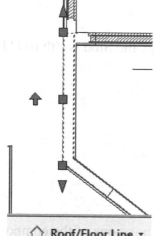

Zoom into one of the walls that was placed. Note that it is the correct width.

The blue arrow indicates the exterior side of the wall. If the blue arrow is inside the building, click on the blue arrow and it will flip the orientation of the wall. Because these are interior walls with gypsum board on both sides, the orientation doesn't matter.

If necessary, move walls so they are aligned with the floor plan's walls.

18.

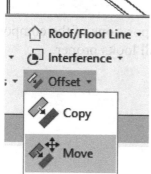

Use the OFFSET tool to move the interior walls to the correct position.

19.

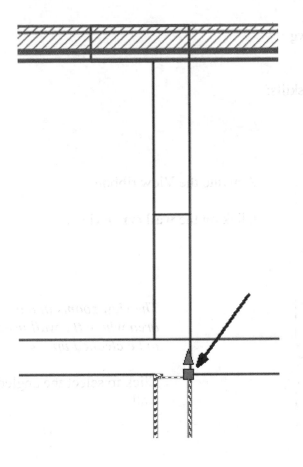

To extend a wall:

Click on the wall to select.

Select the end grip.

20.

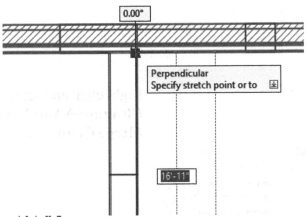

Select a perpendicular OSNAP to connect to an existing wall.

Hint: Turn ORTHO on to keep walls straight.

21.  Some of your walls may display a warning symbol.

This means that you have walls overlapping each other.

Check to see if you have more than one wall or if you need to trim the walls.

22. Save as *ex3-3.dwg*.

Exercise 3-4:

Wall Cleanup

Drawing Name: wall_cleanup.dwg
Estimated Time: 10 minutes

This exercise reinforces the following skills:

- ❑ Using Wall Cleanup
- ❑ Wall Properties

1.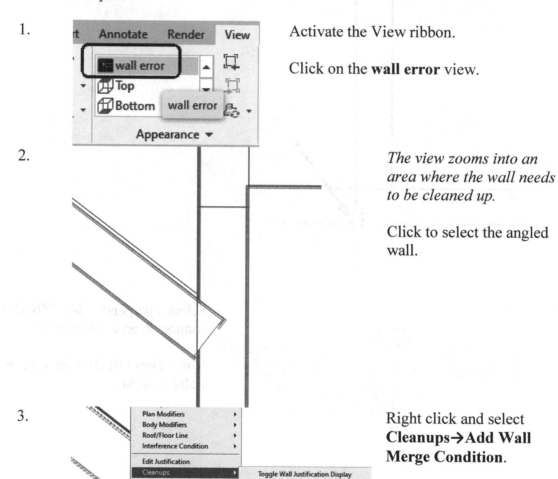

   Activate the View ribbon.

   Click on the **wall error** view.

2. *The view zooms into an area where the wall needs to be cleaned up.*

   Click to select the angled wall.

3. Right click and select **Cleanups→Add Wall Merge Condition**.

4.

Select the vertical wall.

Click **ENTER**.

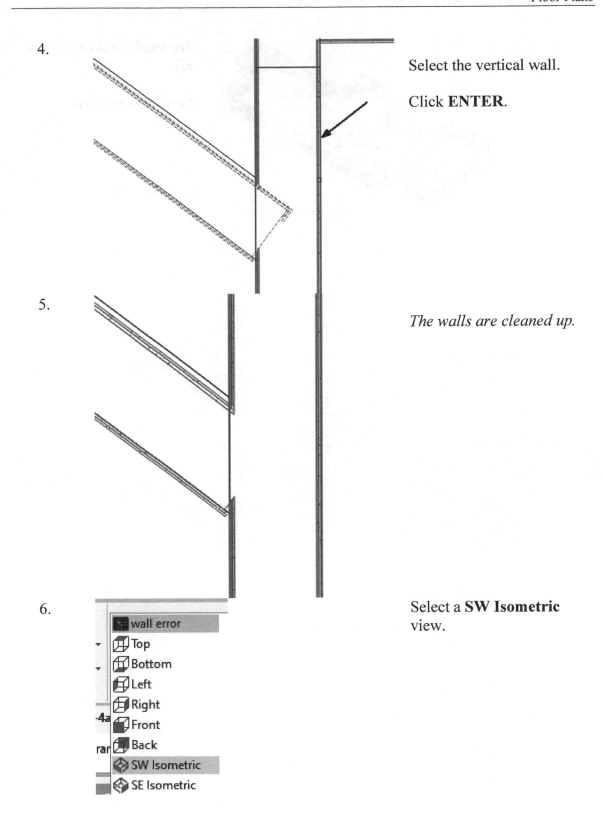

5.

The walls are cleaned up.

6.

Select a **SW Isometric** view.

7.

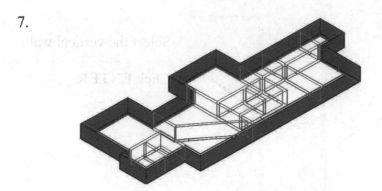

The model is displayed as 3D.

Save as *ex3-4.dwg*.

Exercise 3-5:

Importing a PDF into ACA

Drawing Name: New, floorplan.pdf
Estimated Time: 10 minutes

This exercise reinforces the following skills:

- Import PDF
- Create Walls
- Wall Properties
- Wall Styles
- Model and Workspace

1.   Go to the Start tab.

      Select **New→Aec Model(Imperial Ctb).dwt**.

      This starts the new drawing using imperial units.

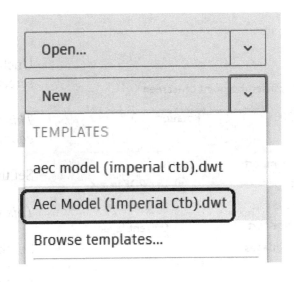

2.   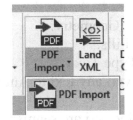 Activate the **Insert** tab on the ribbon.

      Select the **Import PDF** tool (located in the middle of the tab on the ribbon).

3.

File name:	floorplan.pdf
Files of type:	PDF (*.pdf)

Select the *floorplan.pdf* file.

4.

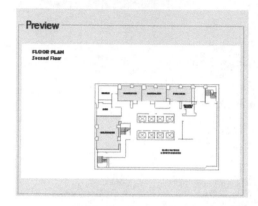

Preview

FLOOR PLAN
Second Floor

You will see a preview of the pdf file.

Click **Open**.

Location

☐ Specify insertion point on-screen

Scale: 1200   Rotation: 0 ⌄

PDF data to import
☑ Vector geometry
☑ Solid fills
☑ TrueType text
☐ Raster images

Layers
◉ Use PDF layers
○ Create object layers
○ Current layer

Import options
☐ Import as block
☑ Join line and arc segments
☑ Convert solid fills to hatches
☑ Apply lineweight properties
☐ Infer linetypes from collinear dashes

Uncheck specify insertion point on-screen.

This will insert the pdf to the 0,0 coordinate.

Set the Scale to **1200**.

This will scale the pdf.

Set the rotation to 0.

Enable Vector Geometry.
Enable Solid Fills.
Enable TrueType Text.
This will convert any text to AutoCAD text.
Enable Join line and arc segments.
Enable Convert solid fills to hatches.
Enable Apply lineweight properties.
Enable Use PDF layers.

5. Click **OK**.

6.

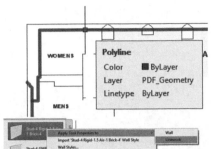

Notice if you hover your mouse over any of the elements imported, they have been converted to ACA elements.

7.

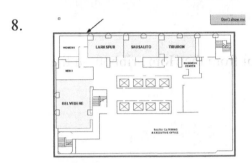

Highlight the **Stud-4 Rigid** wall style on the tool palette. Right click and select **Apply Tool Properties to Linework.**

8.

Select the outside polyline on the floorplan.
Click ENTER to complete the selection.

When prompted to erase existing lines, select **No**.

9.

Highlight the **Stud 4- GWB** wall style.
Right click and select **Apply Tool Properties to Linework.**

Select the lines used for the interior walls.
Click ENTER to complete the selection.

When prompted to erase existing lines, select **No**.

10.

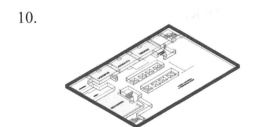

Use the OFFSET and FILLET tools to clean up the floor plan.

11. Save as *ex3-5.dwg*.

You can compare your model with mine and see how you did.

Exercise 3-6:
Creating Walls

Drawing Name: wall_space
Estimated Time: 10 minutes

This exercise reinforces the following skills:

- ❑ Create Walls
- ❑ Wall Properties
- ❑ Wall Styles
- ❑ Model and Workspace
- ❑ Content Browser

1. 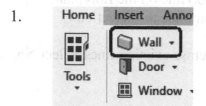 Select the **Wall** tool from the Home ribbon on the ribbon.

2. In the Properties dialog, check under the Style drop-down list.

Style	Standard
Bound spaces	Brick_Block
Cleanup automatically	CMU-8 Rigid-1.5 Air-2 Brick-4
Cleanup group definition	Standard

   These are the styles currently available for use loaded in the drawing.

3. Select the **CMU-8 Rigid-1.5 Air 2 Brick-4** wall style.

Style	CMU-8 Rigid-1.5 Air-2 Brick-4
Bound spaces	By style (Yes)
Cleanup automatically	Yes

4.

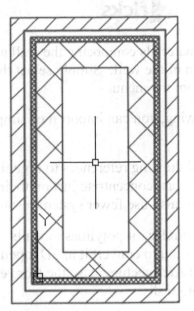

Toggle **ORTHO** ON.

Start the wall at 0,0.
Create a rectangle 72 inches
[1830 mm] tall and 36
inches [914 mm] wide.

*You can use Close to close
the rectangle.*

5. Go to the **View** ribbon.

6. Toggle on the Layout tabs.

7. Select the **Work** tab visible in the lower left
corner of the screen.

8.

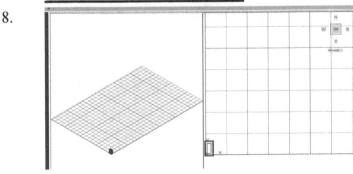

The work tab opens up a
layout with two viewports.
One viewport is 3D and the
other viewport is a top
view.

You see that the walls you
placed are really 3-
dimensional.

9. Save your drawing as *ex3-6.dwg*.

➢ If you draw a wall and the materials composing the wall are on the wrong side, you can reverse the direction of the wall. Simply select the wall, right click and select the Reverse option from the menu.

➢ To add a wall style to a drawing, you can import it or simply create the wall using the Design Tools.

➢ Many architects use external drawing references to organize their projects. That way, teams of architects can concentrate just on their portions of a building. External references also use fewer system resources.

➢ You can convert lines, arcs, circles, or polylines to walls. If you have created a floor plan in AutoCAD and want to convert it to 3D, open the floor plan drawing inside of AutoCAD Architecture. Use the Convert to Walls tool to transform your floor plan into walls.

➢ To create a freestanding door, click the ENTER key when prompted to pick a wall. You can then use the grips on the door entity to move and place the door wherever you like.

➢ To move a door along a wall, use Door→Reposition→Along Wall. Use the OSNAP From option to locate a door a specific distance from an adjoining wall.

Exercise 3-7:

Inserting an Image

Drawing Name: insert_image.dwg
Estimated Time: 10 minutes

This exercise reinforces the following skills:

- ❏ Insert Image
- ❏ Properties
- ❏ Lock Layer

1. 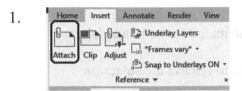 Select the **Insert** tab on the ribbon.

   Select the **Attach** tool.

2. Browse to the folder where the exercises are downloaded.

   Change the Files of type to **All image files**.

3. Select the *floorplan1* file.

   Click **Open**.

4. Uncheck the insertion point to insert the image at **0, 0, 0**.

   Set the Scale to **113.00**.
   Set the Angle to **0.0**.

   Click **OK**.

5.

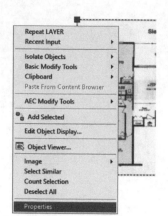

Select the image.
Right click and select **Properties**.

6.

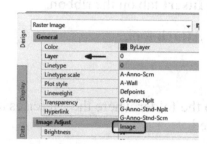

Assign the image to the image layer.

Click ESC to release the selection.

7.

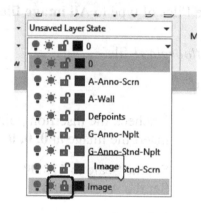

Lock the image layer.

8.

Save as *ex3-7.dwg*.

Exercise 3-8:

Creating a Floorplan from an Image

Drawing Name: image_floorplan.dwg
Estimated Time: 60 minutes

This exercise reinforces the following skills:

- ❑ Wall
- ❑ Wall Properties
- ❑ Wall Style
- ❑ Offset
- ❑ Quick Measure

1.      Select the **Wall** tool from the Home ribbon.

2.   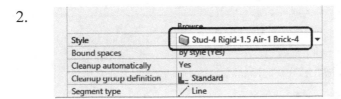   Select the **Stud-4 Rigid-1.5 Air-1 Brick-4** wall style from the Style drop-down list on the Properties palette.

This wall style was pre-loaded in the drawing.

3.   

Turn **ORTHO** on.

Draw a wall on the far left side of the floor plan, tracing over the wall shown in the image file.

Orient the wall so the exterior side of the wall is on the outside of the building.

4.
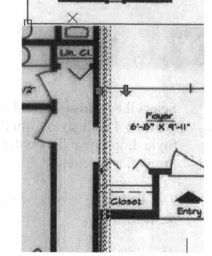

Offset the wall 15′ 11-1/8″ to the right.

Type **OFFSET**.
Type **15′ 11-1/8″** for the offset distance.
Select the wall.
Click to the right of the wall to indicate the side to place the new wall.

The additional offset takes into account the wall thickness of 11-1/8″.

5.

Flip the wall orientation so the wall exterior is on the outside of the building.

6.

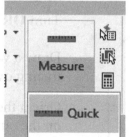

Use the Quick Measure tool to check the offset distance to ensure the two walls are 15′ apart from inside finish face to inside finish face.

7.

Select the **Wall** tool from the Home ribbon.

8.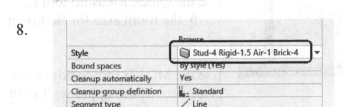

Select the **Stud-4 Rigid-1.5 Air-1 Brick-4** wall style from the Style drop-down list on the Properties palette.

9.

Draw a wall at the bottom of the Bedroom #2 area, connecting the two vertical walls.

Hint: If you draw the wall from right to left, then it will be oriented with the exterior wall on the outside.

10.

Offset the bottom horizontal wall 12' 4.625" above.

Type **OFFSET**.
Type **12'-4.625"** for the offset distance.
Select the wall.
Click above the wall to indicate the side to place the new wall.

The additional offset takes into account the wall thickness of 11-1/8".

11.

Select the upper horizontal wall that was just placed. Right click and select **Properties**.

12.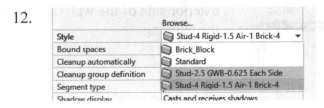

Change the wall style to **Stud-4 GWB-0.625 Each Side.**

Click **ESC** to release the selection.

13.

Use the Quick Measure tool to verify the room size for Bedroom #2.

Use the MOVE tool to adjust the placement of the upper wall so the distance finish face to finish face is 11' 5 ½".

14.

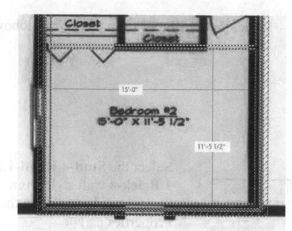

Select the **Wall** tool from the Home ribbon.

15.

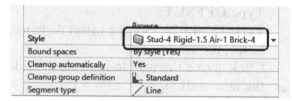

Select the **Stud-4 Rigid-1.5 Air-1 Brick-4** wall style from the Style drop-down list on the Properties palette.

16.

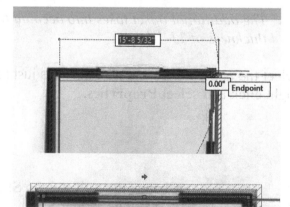

Draw a wall from left to right at the top of the vertical walls.

17.

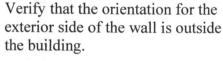

Verify that the orientation for the exterior side of the wall is outside the building.

18.

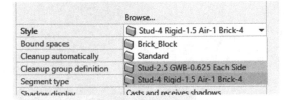

Offset the top horizontal wall 16' 10.625".

Type **OFFSET**.
Type **16'-10.625"** for the offset distance.
Select the wall.
Click below the wall to indicate the side to place the new wall.

19.

Select the horizontal wall that was just placed. Right click and select **Properties**.

20.

Change the wall style to **Stud-2.5 GWB-0.625 Each Side.**

Click **ESC** to release the selection.

21.

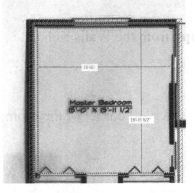

Use the Quick Measure tool to verify the Master Bedroom dimensions.

MOVE the top horizontal wall as needed to get the correct dimensions.

22.

Use an offset of 2' 9" to create the closet space for the master bedroom.

Quick Measure will display a finish face to finish face dimension of 2' 5.25"

23.

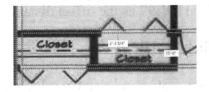

Use an offset of 2' 7.5" to create the closet space between the bedroom #3 and bedroom #2.

Quick Measure will display a finish face to finish face dimension of 2' 3.75"

24.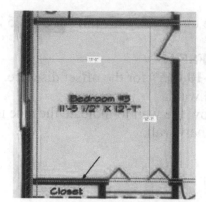

Offset the upper closet wall at the bottom of Bedroom #3 12'-10.75".

Quick Measure should show a vertical distance of 12'- 7".

25.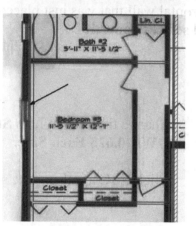

Select the left vertical wall for Bedroom #3.

26.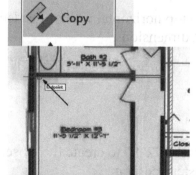

Select **Offset→Copy** from the Walls ribbon.

27.

Select the upper left corner of the room.

28.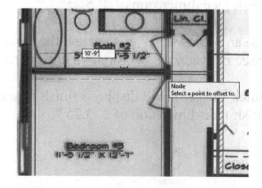

Move the mouse slightly to the right. Type **10'-9"** as the offset distance.

Click ENTER.

Click ESC to release the selection and exit the command.

29.

Select Similar

Count Selection

Deselect All

Properties

Select the wall that was just placed.
Right click and select **Properties**.

30.

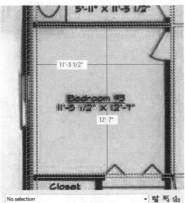

Change the wall style to **Stud-2.5 GWB-0.625 Each Side.**

Click **ESC** to release the selection.

31.

Use Quick Measure to check the dimensions for Bedroom #3.

Move the interior wall as necessary.

32.

No selection		
Style	Stud-2.5 GWB-0.625 Each Side	
Bound spaces	By style (Yes)	
Cleanup automatically	Yes	
Cleanup group definition	Standard	
Segment type	Line	
Dimensions		
A Width	3 3/4"	
B Base height	10'-0"	
C Length	1"	
Justify	Center	
Offset	0"	
*E Roof line offset from base...	0"	
*F Floor line offset from base...	0"	

Trace the remaining walls of the floor plan.

Set Justify to **Center** in the Properties palette to line the walls up to the image lines.

33.

As you work, use the Quick Measure tool to verify the wall placement.

34.

Save as *ex3-8.dwg*.

Exercise 3-9:
Controlling Image Visibility

Drawing Name: image_properties.dwg
Estimated Time: 5 minutes

This exercise reinforces the following skills:

- Images
- Layers
- IMAGEFRAME

1. 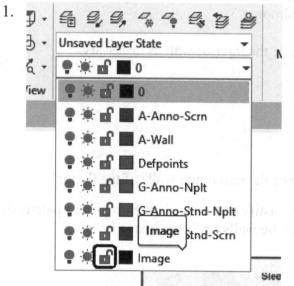 Unlock the image layer.

2. 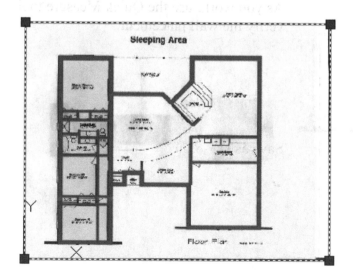 Select the image.

3. On the Image ribbon:
On the Adjust panel.
You can adjust how much of the image you see so it doesn't interfere with your work.

Alternatively, you can freeze the image layer or change the transparency of the layer.

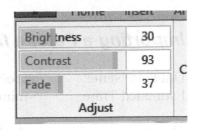

You can also toggle the Show Image tool to turn the visibility of the image on or off.

To turn off the image border, type **IMAGEFRAME** and set the value to **0**.

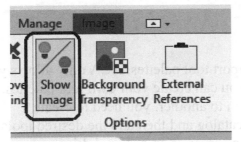

4.

You should have a completed floor plan.

Save as *ex3-9.dwg*.

Exercise 3-10:

Importing a Palette from the Content Browser

Drawing Name: import_palette.dwg
Estimated Time: 5 minutes

This exercise reinforces the following skills:

- ❑ Content Browser
- ❑ Tool Palettes

While AutoCAD will allow you to export and import tool palettes, ACA only allows you to add tool palettes using the Content Browser. You can create your own tool palettes. In order to transfer tool palettes from one workstation to another, you need to create a custom catalog in the Content Browser, create a catalog and then add the desired tools to the catalog. You can store the custom catalog on a server or on a shared drive.

1. 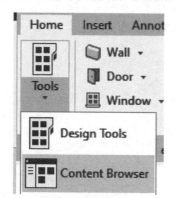 Switch to the Home ribbon.

   Go **Tools→Content Browser**.

2.  In the lower left of the dialog window:

   Click **Create or Add Catalog**.

3. 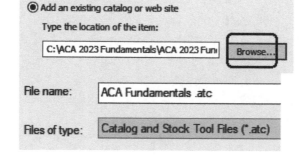 Enable **Add an existing catalog or website**.

   Click **Browse**.

4. Browse to the downloaded files and locate the *ACA Fundamentals.atc* file.

   Click **Open**.

   Click **OK**.

5.

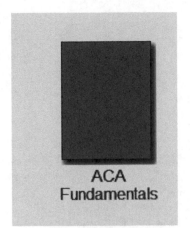

The catalog will be added to your Content Browser.

Click on the catalog.

6.

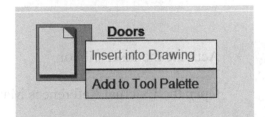

You see a Doors Category.

Right click and select **Add to Tool Palette**.

7.

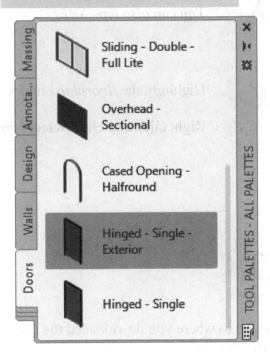

The tool palettes open with the new imported Doors palette.

Exercise 3-11:
Adding Doors

Drawing Name: add_doors.dwg
Estimated Time: 45 minutes

This exercise reinforces the following skills:

- ❑ External References
- ❑ Adding Doors
- ❑ Door Properties

1. 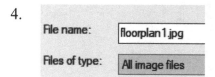 Open *add_doors.dwg*.

2. 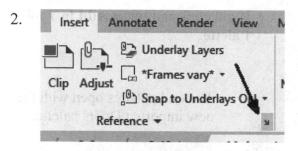 Activate the Insert ribbon.

   Open the External References Manager.

   You can also type XREF.

3. 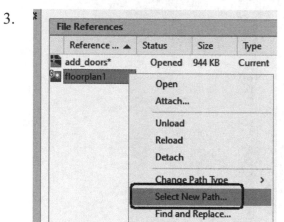 Highlight the *floorplan1* image file.

   Right click and click **Select New Path**.

4.
File name:	floorplan1.jpg
Files of type:	All image files

   Browse to where you downloaded the file.
   Select the file.
   Click **Open**.

   Because the drawing is referencing an external file, you need to point it to the correct path.

   Close the XREF Manager.

5. Type **PLAN** to switch to a plan view.
   Click **ENTER**.

6. 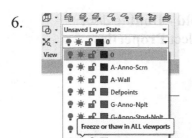 Switch to the Home ribbon.

Thaw the image layer so you can see where doors are located.

7. 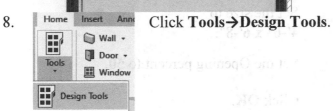 Zoom into the Closets between the two bedrooms.

8. 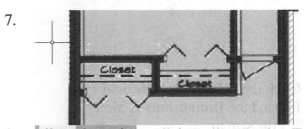 Click **Tools→Design Tools**.

9. 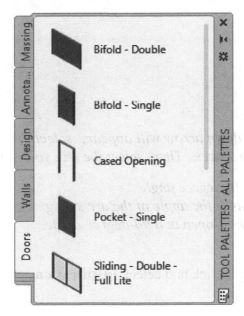 Select the **Doors** tab on the palette.

10.

Highlight the **Bifold - Double** door.
Right click and select **Properties**.

11.

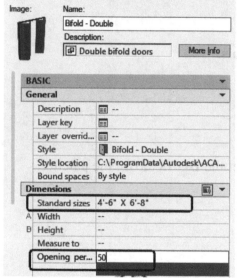

Click the **Bifold - Double** door tool.
Expand the **Dimensions** section.
Set the size to
4'-6" x 6'-8".

Set the Opening percent to **50**.

Click **OK**.

If you left click in the Standard sizes field, a down arrow will appear...select the down arrow and you will get a list of standard sizes. Then, select the size you want.

A 25% opening will show a door swing at a 45-degree angle.
The value of the Opening percentage determines the angle of the arc swing.
A 50% value indicates the door will appear half-open at a 90-degree angle.

12.

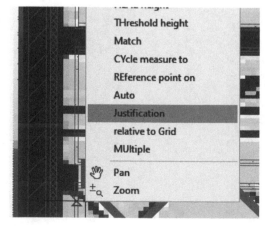

Right click and select **Justification**.

13. 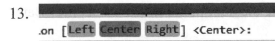 Select **Center**.

.on [Left Center Right] <Center>:

14. Place the Bifold - Double doors at the two closets.

The orientation of the door swing is determined by the wall side selected.

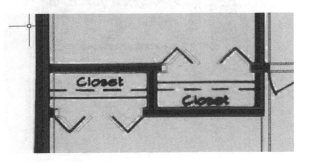

In both cases, you want to select the outside face of the wall. Center the closet door on each wall.

15. 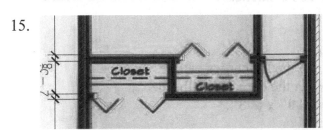 Place the **Bifold - Double** door at each of the closets located in Bedroom #2 and Bedroom #3.

16. 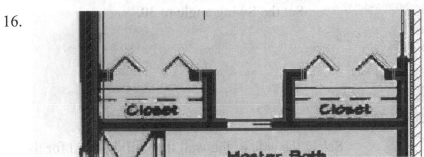 Place the **Bifold - Double** door at the closets in the Master Bedroom.

17. Bifold - Single Click the **Bifold - Single** door on the Doors tab of the Design Tools palette.

18. 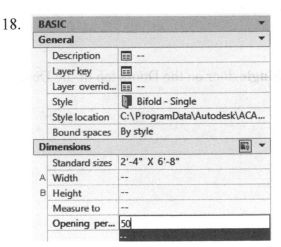 In the Properties palette:

Set the door to use the Standard Size **2′ 4″ x 6′ 8″**.
Set the Opening percent to **50**.

19. 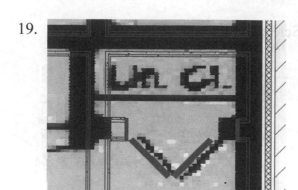 Place the door in the Linen Closet near Bath #2.

20.  Hinged - Single - Exterior

Select the **Hinged - Single - Exterior** door on the Doors tab of the Design Tools palette.

21.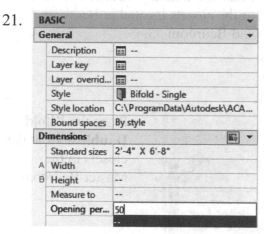

BASIC		
General		
Description	⊞ --	
Layer key	⊞	
Layer overrid...	⊞ --	
Style	Bifold - Single	
Style location	C:\ProgramData\Autodesk\ACA...	
Bound spaces	By style	
Dimensions		
Standard sizes	2'-4" X 6'-8"	
A Width	--	
B Height	--	
Measure to	--	
Opening per...	50	
	--	

In the Properties palette, set the door to use the size **3′ 0″ x 6′ 8″**.

Set the Swing angle to **30**.

22.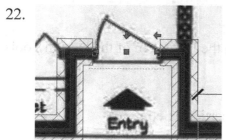

Select the side of the wall that will be used for the door swing and place the entry door.

23.  Hinged - Single

Select the **Hinged - Single** door on the Doors tab of the Design Tools palette.

24.

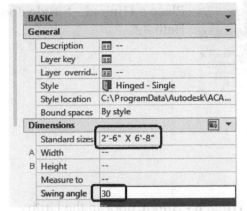

In the Properties palette, set the door to use the size **2′ 6″ x 6′ 8″**.

Set the Swing angle to **30**.

25.

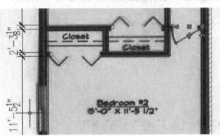

Place the door in Bedroom #2.

26.

Place the door in Bedroom #3.

The swing is on the correct side but not the correct direction.

27.

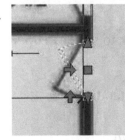

Select the door so it highlights.

The horizontal arrow flips the orientation of the door to the other side of the wall.

The vertical arrow flips the orientation of the door swing.

Left click on the vertical arrow.

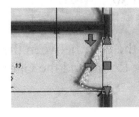

The door updates to match the floor plan image.

28. 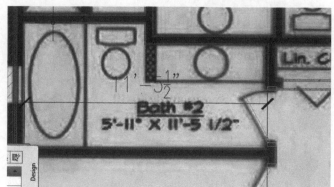 Place a **Hinged - Single** door in Bath #2.

29. 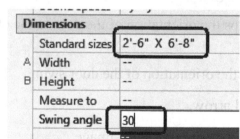 Place a **Hinged - Single** door in the Utility Room.

30. 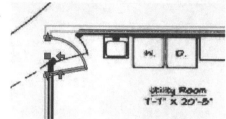 Change the size to **3'-0" x 6'-8"**.

Set the swing angle to **80**.

31. ![Hinged - Single - Exterior] Locate the **Hinged - Single - Exterior** door on the Doors tab of the Design Tools palette.

32.

Dimensions	
Standard sizes	2'-6" X 6'-8"
A Width	--
B Height	--
Measure to	--
Swing angle	30

Highlight and right click to select **Properties.**

In the Properties palette,
set the door to use the Standard Size
2' 6" x 6' 8".

Set the Swing angle to **30**.

33. 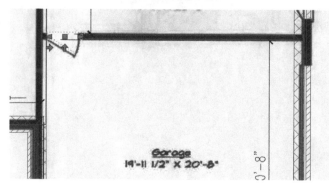 Place the door between the Utility Room and the Garage.

34. 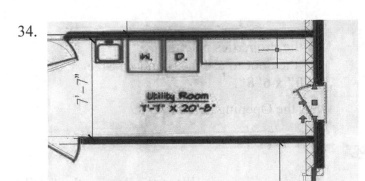 Place the door on the east wall of the Utility Room.

35. **Overhead - Sectional** — Locate the **Overhead - Sectional** door on the Doors tab of the Design Tools palette.

36.

Dimensions	
Standard sizes	16'-0" X 6'-8" (Custom Size)
A Width	--
B Height	6'-8"
Measure to	--
Opening percent	0

Highlight and right click to select **Properties.**

In the Properties palette, set the door to use the Size **16′ 0″ x 6′ 8″.**

To set the size, select the Custom Size option in the drop-down list. Type in the desired width and height. The size will adjust.

Set the Opening percent to **0.**

37. Place the garage door.

Garage
19'-11 1/2" X 20'-8"
19'−11½"
20'−8"

38. **Sliding - Double - Full Lite** — Select the **Sliding - Double - Full Lite** door on the Doors tab of the Design Tools palette.

39.

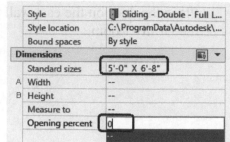

In the Properties palette,
set the door to use the Standard Size
5′ 0″ x 6′ 8″.

Set the Opening percent to **0**.

40.

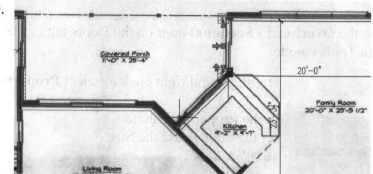

Place the door in the family room.

41.

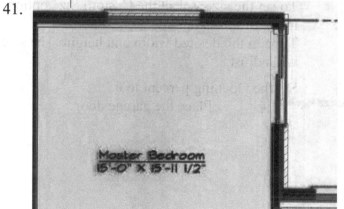

Place a second **Sliding - Double - Full Lite** door on the east wall of the Master Bedroom.

42.

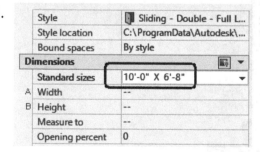

Set the door to use the size
10′ 0″ x 6′ 8″.
Set the Opening Percent to **0**.

43. 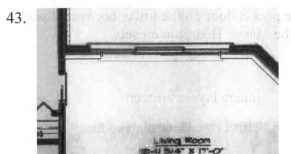 Center the door on the north wall of the Living Room.

44.  Select the **Pocket - Single** door on the Doors tab of the Design Tools palette.

45. 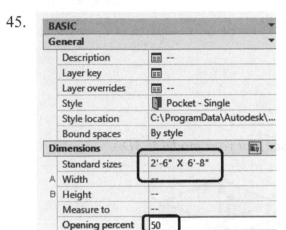 In the Properties palette set the door to use the size: **2′ 6″ x 6′ 8″**.

Set the Opening percent to **50**.

46. 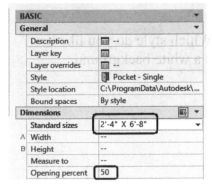 Place the door in the lower right corner of the Master Bedroom.

47.  Select the **Pocket - Single** door on the Doors tab of the Design Tools palette.

48. In the Properties palette, set the door to use the Standard Size **2′ 4″ x 6′ 8″**.

Set the Opening percent to **50**.

49. 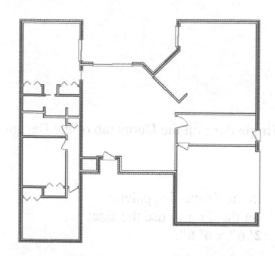 Center the pocket door on the lower horizontal wall between the Master Bedroom closets.

50.

Image layer is frozen.

This is the floor plan so far.

51. Save as *ex3-11.dwg*.

Switch to an isometric view and you will see that your model is 3D.

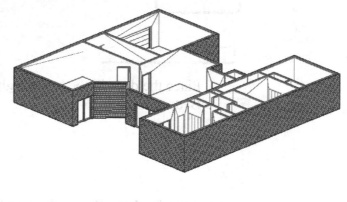

Look at the model using different visual styles. Which style do you like best? The model shown uses a Hidden visual style with a white background.

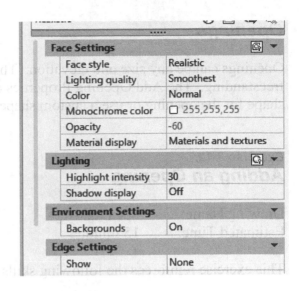

This is a Realistic style.

Openings

Openings can be any size and elevation. They can be applied to a wall or be freestanding. The Add Opening Properties allow the user to either select a pre-defined shape for the opening or use a custom shape.

Exercise 3-12:
Adding an Opening

Drawing Name: opening.dwg
Estimated Time: 15 minutes

This exercise reinforces the following skills:

- ❑ Adding Openings
- ❑ Opening Properties
- ❑ Copying Tools
- ❑ Set Image from Selection

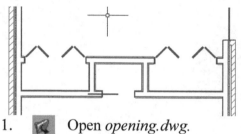

An opening will be added to the upper wall between the Master Bedroom closets.

1. 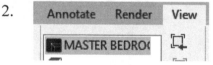 Open *opening.dwg*.

2. Go to the View ribbon.

   Activate the **MASTER BEDROOM OPENING** view.

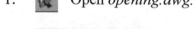

The view updates to the location where the opening will be placed.

3. 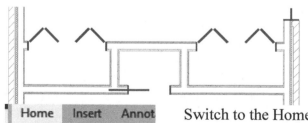 Switch to the Home ribbon.

   Click **Tools→Design Tools**.

4.

✓	Design
	All Palettes

Right click on the Tool Palettes bar.
Enable **All Palettes**.

5.

Cased Opening -
Halfround

Select the **Cased Opening – Halfround** tool on the
Doors palette.

6.

Style	Cased Opening - Halfround
Bound spaces	By style (Yes)
Dimensions	
Standard sizes	2'-6" X 6'-8"
A Width	2'-6"
B Height	6'-8"
Measure to	Inside of frame
Opening percent	50

In the Properties palette,
set the door to use the size
2' 6" x 6' 8".

7.

Location	
✱ Relative to grid	No
✱ Position along wall	Offset/Center
✱ Automatic offset	6"
✱ Justification	Center
✱ Multiple insert	No
Vertical alignment	Threshold

Expand the Location section in the Properties
palette.
Set the Position along wall to **Offset/Center**.
Set the Automatic offset to **6" [300.00]**.

8.

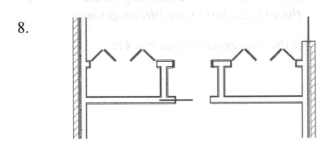

Place the arched opening in the wall
between the closets in the Master
Bedroom. Center it on the wall.

9.

Dimensions	
Standard sizes	3'-0" X 6'-8"
A Width	3'-0"
B Height	6'-8"
Measure to	Inside of frame
Opening percent	50

In the Properties palette,
set the door to use the size
3' 0" x 6' 8".

10.

Place the Arched Opening on the left side of the Foyer above the Entry.

11. Use the View tools on the View tab on the ribbon **View → NE Isometric** and **3D orbit** to view the arched opening.

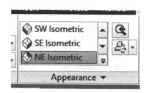

12.

On the View tab on the ribbon,

Switch to a **Shades of Gray** display.

If your walls are reversed, you can change the orientation in the plan/top view.

13.

Set the **Materials/Textures On**.

14.

Set to **Full Shadows**.

Note how the display changes.

When materials, textures, and shadows are enabled, more memory resources are used.

15. Select the Work tab to view your model.

16.

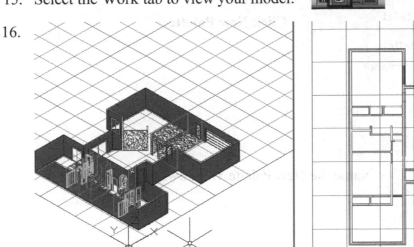

17. Save the file as *ex3-12.dwg*.

Exercise 3-13:

Create A Tool Palette using the Styles Manager

Drawing Name: windows_adc.dwg
Estimated Time: 30 minutes

This exercise reinforces the following skills:

- ❑ Styles Manager
- ❑ Tool Palettes
- ❑ New Palette

1.   Activate the **Manage** ribbon.

     Click **Style Manager**.

2.    Go **Tools→Design Tools**.

3.

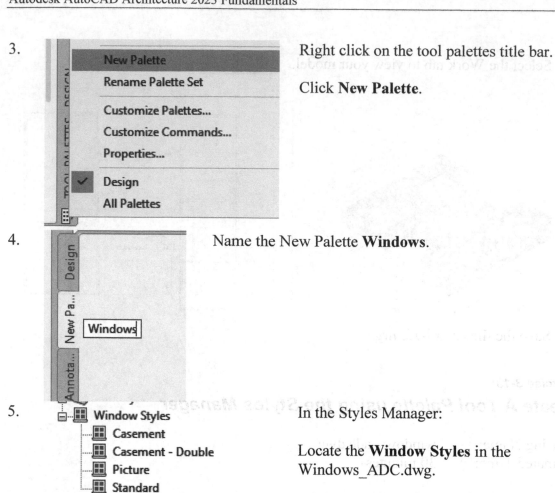

Right click on the tool palettes title bar.

Click **New Palette**.

4. Name the New Palette **Windows**.

5. In the Styles Manager:

Locate the **Window Styles** in the Windows_ADC.dwg.

6. Drag and drop all the window styles from the Styles Manager onto the Windows tool palette.

Close the Styles Manager.

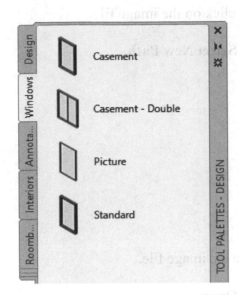

You now have a Windows tool palette.

Exercise 3-14:
Add Windows

Drawing Name: windows.dwg
Estimated Time: 30 minutes

This exercise reinforces the following skills:

☐ Add Windows

1.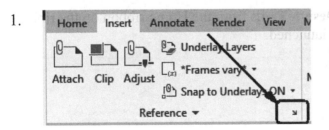

Switch to the **Insert** ribbon.

Launch the **XREF Manager.**

2. 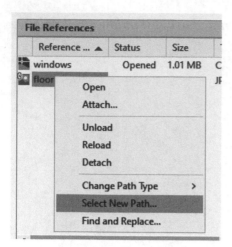 Right click on the image file.

Click **Select New Path**.

3. 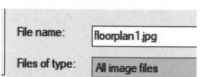 Locate the image file.

Click **Open**.

4. Close the **XREF Manager**

5. 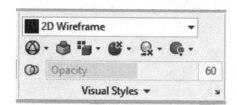 Set the View style to **2D Wireframe**.

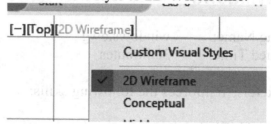

Remember you can change the view settings in the upper left corner of the display window.

6. 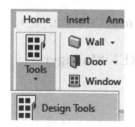 Activate the **Design Tools** from the Home ribbon on the ribbon, if they are not launched.

7. Switch to the Home ribbon.

Thaw the image layer, so you can see the location of the windows.

8.

Switch to a PLAN view.
Type **PLAN, ENTER**.

9. Select the **Windows** tab of the Tool palette.

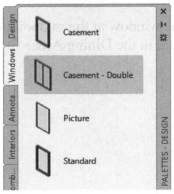

10. Select the **Picture** window.

11.

Dimensions	
Standard sizes	6'-0" X 5'-0"
A Width	6'-0"
B Height	5'-0"
Measure to	Outside of frame
Opening percent	0

Expand the Dimensions section.
Set the size to **6′-0″ x 5′-0″**.

12.

Location	
❈ Relative to grid	No
❈ Position along wall	Offset/Center
❈ Automatic offset	6"
❈ Justification	Center
❈ Multiple insert	No
Vertical alignment	Head
Head height	6'-8"
Sill height	1'-8"
Rotation	0.00

Expand the Location section.
Set the Position to **Offset/Center**.
Set the Automatic Offset to **6″**.

13.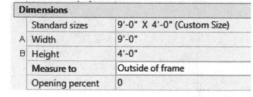

Select the midpoint of the north Master Bedroom wall.

14.

Dimensions	
Standard sizes	9'-0" X 4'-0" (Custom Size)
A Width	9'-0"
B Height	4'-0"
Measure to	Outside of frame
Opening percent	0

On the Properties palette,
expand the Dimensions section.

Change the Width to **9′-0″**.
Change the Height to **4′-0″**.

15.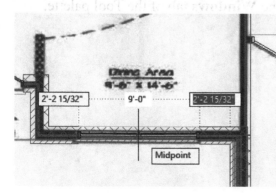

Place the window at the midpoint of the south wall in the Dining Area.

16.

Dimensions	
Standard sizes	12'-10" X 4'-0" (Custom Size)
A Width	12'-10"
B Height	4'-0"
Measure to	Outside of frame
Opening percent	0

On the Properties palette, expand the Dimensions section.

Change the Width to **12'-10"**.
Change the Height to **4'-0"**.

17.

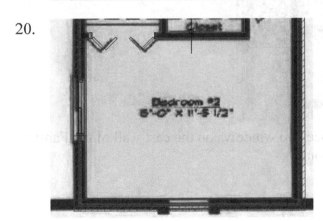

Place the window at the midpoint of the north wall for the Family Room.

18. Casement - Double

Select the **Casement - Double** window on the Windows palette.

19.

Dimensions	
Standard sizes	4'-0" X 4'-0"
A Width	4'-0"
B Height	4'-0"
Measure to	Outside of frame
Swing angle	0

On the Properties palette, expand the Dimensions section.

Set the Standard size to **4'-0" x 4'-0"**.

20.

Place the window in the west wall of Bedroom #2.

21. 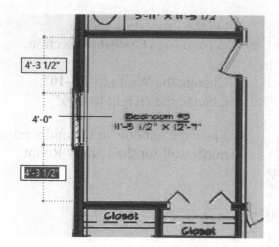 Place the window in the west wall of Bedroom #3.

22. 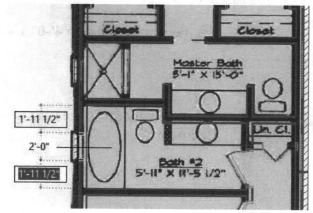 Select the **Casement** window.

23.

Dimensions	
Standard sizes	2'-0" X 4'-0" (Custom Size)
A Width	2'-0"
B Height	4'-0"
Measure to	Outside of frame
Swing angle	0

On the Properties palette, expand the Dimensions section.

Change the Width to **2′-0″** .
Change the Height to **4′-0″**.

24.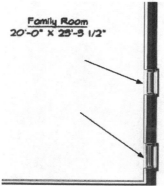

Place two windows on the west wall of the bathrooms.

25.

Place two windows on the east wall of the Family Room.

26.

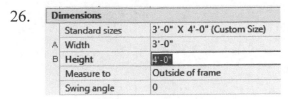

On the Properties palette,
expand the Dimensions section.

Change the Width to **3′-0″** .
Change the Height to **4′-0″**.

27.

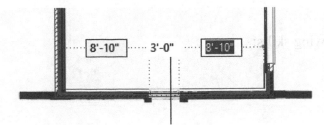

Place the window in the
south wall of the
Garage.

28. Save as *ex3-14.dwg.*

Grids

Grids are used in floor plans to help architects and designers locate AEC elements, such
as columns, walls, windows, and doors. The grid is one of the oldest architectural design
tools – dating back to the Greeks. It is a useful tool for controlling the position of
building elements. Grids may fill the entire design work area or just a small area.

In laying out wood framed buildings, a 16" or 24" grid is useful because that system
relies on a stud spacing of 16" or 24".

In large projects, the layout design of different building sub-systems and services are
usually assigned to different teams. By setting up an initial set of rules on how the grid is
to be laid out, designers can proceed independently and use the grid to ensure that the
elements will interact properly.

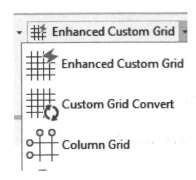

ACA comes with several grid tools.

The Enhanced Custom Grid allows you to place an entire grid
using a dialog box to designate the bay spacing.

The Custom Grid Convert tool allows you to convert lines and
arcs to grid lines. After the grid lines are placed, you have to
manually edit the labels for each bubble.

The Column Grid tool allows you to place a rectangular grid with
equal spacing between the grid lines.

Use the Enhanced Custom Grid tool if you want to use different
spacing between grid lines or if you want to place radial grid
lines.

Exercise 3-15:
Place a Grid

Drawing Name: grid.dwg
Estimated Time: 5 minutes

This exercise reinforces the following skills:

□ Place a grid

1.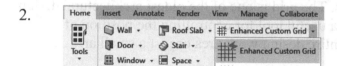

   Open *grid.dwg.*

   Activate the Model tab.
   Switch to a Top View.

2.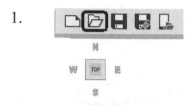

   Click on **Enhanced Custom Grid** on the Home ribbon.

3.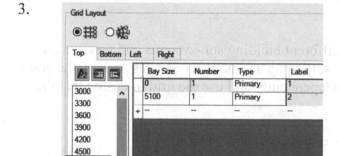

   Click **5100**.

   The first two grids are defined.

   There is a preview window which shows the grids spaced 5100 mm apart.

4.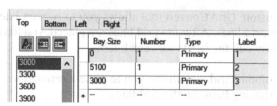

   Click 3000.

   A third grid is placed.

5.

Click **5100**.

A fourth grid is placed.

6.

Switch to the Left tab.

Click **3000**.

The first two grids are defined.

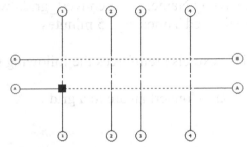

There is a preview window which shows the grids spaced 3000 mm apart horizontally.

7.

Click **3000.**

A third grid is placed.

Notice that because the spacing is set the same, it is listed with the previous grid.

Click **OK** to close the dialog.
Type **0,0** for the insertion point.

8.

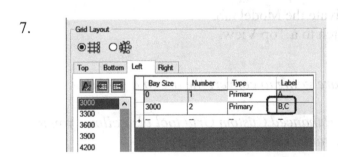

9.

Click **ENTER** to accept the default rotation of 0 degrees.

10.

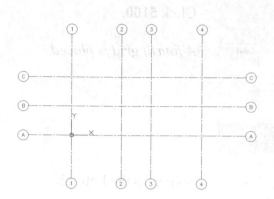

Click ENTER to exit the command.

Double click the wheel of the mouse to Zoom Extents.

The grid is placed.

Save as *ex3-15.dwg*.

Exercise 3-16:
Convert to Grid

Drawing Name: convert_grid.dwg
Estimated Time: 5 minutes

This exercise reinforces the following skills:

□   Convert an arc to a grid

1.

Open *convert_grid.dwg*.

Activate the Model tab.
Switch to a Top View.

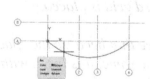

We are going to convert the arc to a grid line.

The Enhanced Custom Grid tool also allows you to place and create radial grids.

2.

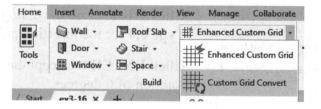

Click **Custom Grid Convert** on the Home ribbon.

3.

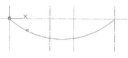

Click on the arc to select.

Click **ENTER** to complete the selection.

4.

Click **ENTER** to accept the default value for the label extensions.

5.

CUSTOMCOLUMNGRIDADD Erase selected linework? [Yes No] <No>:

Click **Yes** to erase the existing linework.

6.

The grid is placed but notice that the labels have not been defined.

Click **ESC** to release the selection.

7.

Select one of the grid bubbles.

Right click and select **Properties**.

8.

In the Properties palette, select **Attributes**.

9.

Change the Value to **D**.
Change the Display to **D**.

Click **OK**.

Click **ESC** to release the selection.

10.

Note that one label updated, but not the other.

Repeat to update the other bubble to also display **D**.

11.

Save as *ex3-16.dwg*.

Exercise 3-17:
Place a Curtain Wall

Drawing Name: curtain_wall.dwg
Estimated Time: 5 minutes

This exercise reinforces the following skills:

❑ Place a curtain wall

1.

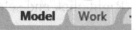

[−][Top][2D Wireframe]

Open *curtain_wall.dwg*.

Activate the Model tab.
Switch to a Top View.

2.

Select the **Curtain Wall** tool on the Home ribbon.

3.

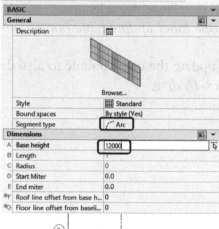

Set the Segment type to **Arc**.

Set the Base height to **12000**.

4.

Select the intersection of A1 as the start point.

5.

Select the Midpoint/Arc Lower Quadrant between Grids 2 and 3 as the midpoint.

6. 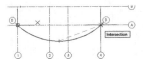 Select the intersection of A4 as the end point.

Click **ENTER** to complete the wall.

7.

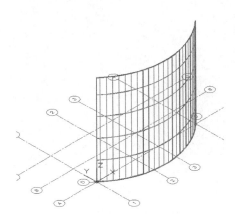

Switch to a SW Isometric view to see the curtain wall.

Save as *ex3-17.dwg*.

Notes:

QUIZ 3

True or False

1. When you insert a PDF into a drawing, it cannot be converted to AutoCAD elements, like lines or text.

2. When you insert an image, it cannot be converted to AutoCAD elements, like lines or text.

3. The direction you place walls – clockwise or counter-clockwise – determines which side of the wall is oriented as exterior.

4. Curtain walls can only be linear, not arcs, in AutoCAD Architecture.

5. Grids can only be lines, not arcs, in AutoCAD Architecture.

6. To re-orient a door, use the Rotate command.

7. You can set cut planes for individual objects (such as windows), an object style (such as walls) or as the system default.

Multiple Choice

8. To change the hatch display of wall components in a wall style:

   A. Modify the wall style
   B. Change the visual style
   C. Change the display style
   D. Switch to a plan\top view

9. This tool on the status bar:

   A. Sets the elevation of the active level
   B. Controls the default cut plane of the view and the display range
   C. Sets the distance between levels
   D. Sets the elevation of the active plane

10. To assign a material to a door:

   A. Modify the door component by editing the door style
   B. Drag and drop the material from the Material Browser onto a door
   C. Use the Display Manager
   D. Define a new Visual Style

ANSWERS:

1) F; 2) T; 3) T; 4) F; 5) F; 6) F; 7) T; 8) A; 9) B; 10) A

Lesson 4:
Space Planning

A residential structure is divided into three basic areas:

- ❑ Bedrooms: Used for sleeping and privacy
- ❑ Common Areas: Used for gathering and entertainment, such as family rooms and living rooms, and dining area
- ❑ Service Areas: Used to perform functions, such as the kitchen, laundry room, garage, and storage areas

When drawing your floor plan, you need to verify that enough space is provided to allow placement of items, such as beds, tables, entertainment equipment, cars, stoves, bathtubs, lavatories, etc.

AutoCAD Architecture comes with Design Content to allow designers to place furniture to test their space.

Exercise 4-1:
Creating AEC Content

Drawing Name: New
Estimated Time: 20 minutes

This lesson reinforces the following skills:

- ❑ Managing External Reference Files
- ❑ XREF Manager
- ❑ Design Center
- ❑ AEC Content
- ❑ Customization

1. 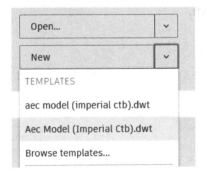 Start a new drawing using the *Aec Model (Imperial.Ctb)* template.

2.

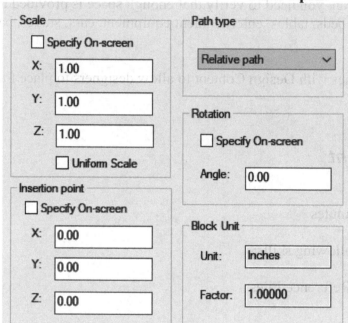

Activate the Insert ribbon.

Select **Attach**.

3.

Browse for *residence.dwg.*
This file was downloaded from the publisher's website.

Click **Open**.

4.

Accept the defaults.

Uncheck **Specify On-screen** for the insertion point.

Click **OK**.

5.

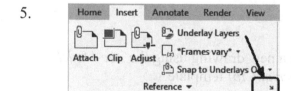

Launch the XREF Manager by selecting the small arrow in the Reference drop-down.

6. The palette lists the file references now loaded in the drawing.
Notice that the floorplan image file is listed as a reference.
The image file is missing because the path link has changed.
The image file is being hosted by the residence.dwg file.

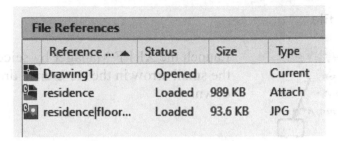

File References			
Reference ... ▲	Status	Size	Type
Drawing1	Opened		Current
residence	Loaded	989 KB	Attach
residence|floor...	Loaded	93.6 KB	JPG

7.

Edit Reference In-Place Open Reference

Click on the model. This is an external reference.

On the ribbon:
Click **Open Reference**.

8. Drawing7* **residence** ✕ +

n View][2D Wireframe]

The residence drawing which is being used as an external reference is opened..

9.

Home Insert Annotate Render View

Attach Clip Adjust

Underlay Layers
Frames vary
Snap to Underlays O

Reference ▾

Launch the XREF Manager by selecting the small arrow in the Reference drop-down.

10.

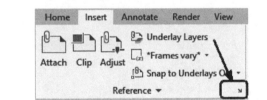

File References

Reference Name ▲

residence

floorplan1

Open
Attach...

Unload
Reload
Detach

Change Path Type ›
Select New Path...
Find and Replace...

Highlight the image file.

Right click and click **Select New Path**.

11.

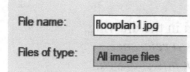

Locate the image file in the downloaded files.

Click **Open**.
Close the XREF Manager.

12. Save the *residence.dwg* file.
Close the file.

13.

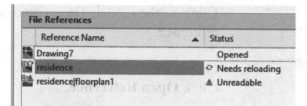

Launch the XREF Manager by selecting the small arrow in the Reference drop-down.

14.

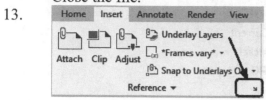

Highlight the *residence* file.

15.

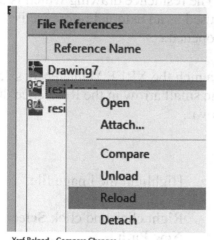

Right click and select **Reload**.

16.

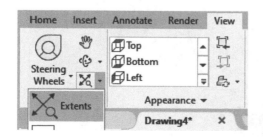

Click **No**.

You know what changed – you updated the path for the image file.
Close the XREF Manager.
Zoom Extents.

You can do this by typing Z,E on the command line or using the tool on the View tab on the ribbon or double clicking the mouse wheel.

17.

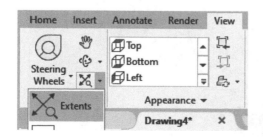

18.

Enable the Nearest OSNAP to make it easier to place furniture.

19.

Launch the Design Center from the Insert tab on the ribbon or Click Ctrl+2 or type **DC**.

20.

Select the **AEC Content** tab.

21.

Browse to *AutoCAD Architecture/Imperial/Design/Furnishing/Furniture/Bed*.

22.  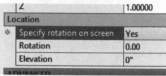 Select the *Double.dwg* file.
Right click and select **Insert**.

23. 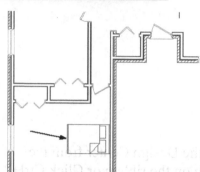 Set Specify rotation to **Yes** in the Properties dialog.

24. 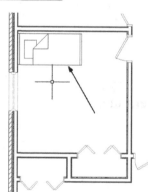 Place the bed in Bedroom #2.

Right click and select ENTER to exit the command.

25. Locate the *Twin.dwg*.

26. 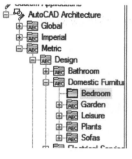 Place the *Twin.dwg* in Bedroom#3.

Right click and select ENTER to exit the command.

27. Browse to the *Bedroom* folder in the *Metric/Design/Domestic Furniture* folder.

28. 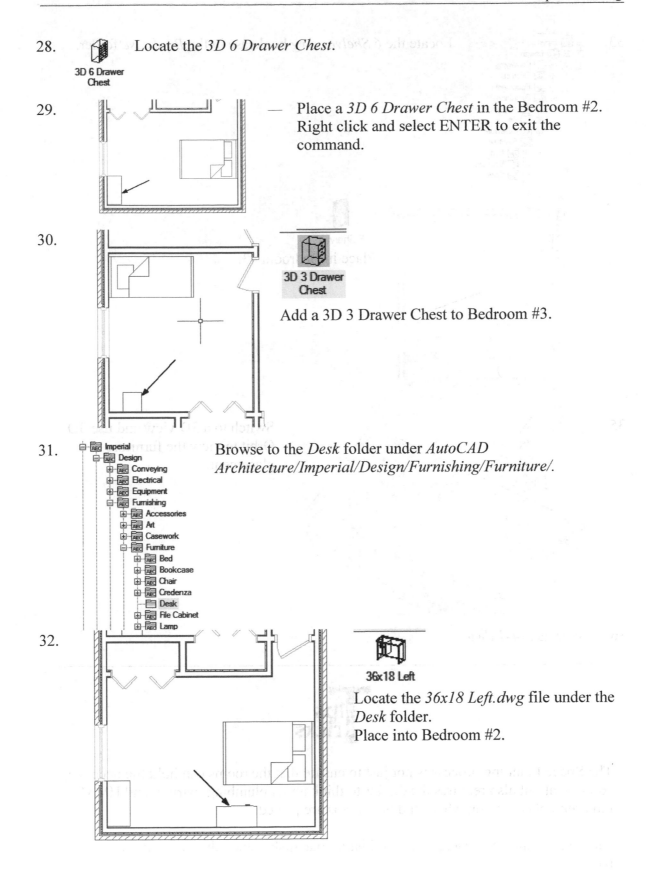 Locate the *3D 6 Drawer Chest*.

3D 6 Drawer
Chest

29. — Place a *3D 6 Drawer Chest* in the Bedroom #2. Right click and select ENTER to exit the command.

30. 3D 3 Drawer
Chest

Add a 3D 3 Drawer Chest to Bedroom #3.

31. Browse to the *Desk* folder under *AutoCAD Architecture/Imperial/Design/Furnishing/Furniture/*.

Imperial
Design
Conveying
Electrical
Equipment
Furnishing
Accessories
Art
Casework
Furniture
Bed
Bookcase
Chair
Credenza
Desk
File Cabinet
Lamp

32. 36x18 Left

Locate the *36x18 Left.dwg* file under the *Desk* folder.
Place into Bedroom #2.

33. 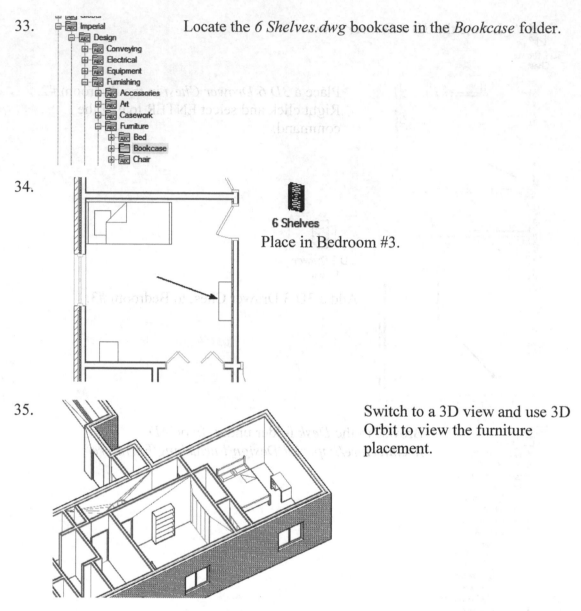 Locate the *6 Shelves.dwg* bookcase in the *Bookcase* folder.

34.

6 Shelves

Place in Bedroom #3.

35. Switch to a 3D view and use 3D Orbit to view the furniture placement.

36. Save as *ex4-1.dwg*.

Tips & Tricks

The Space Planning process is not just to ensure that the rooms can hold the necessary equipment but also requires the drafter to think about plumbing, wiring, and HVAC requirements based on where and how items are placed.

As an additional exercise, place towel bars, soap dishes and other items in the bathrooms.

Exercise 4-2:
Furnishing the Common Areas

Drawing Name: common areas.dwg
Estimated Time: 15 minutes

Common areas are the Living Room, Dining Room, and Family Room.

This lesson reinforces the following skills:

- Insert
- ADC

1. Open *common areas.dwg.*

2. Change to a PLAN view by typing **PLAN** and **ENTER.**.

3.    Type **INSERT** or

    Go to the Insert tab on the ribbon, select
    Insert→Blocks from Libraries.

4. File name: dining group.dwg
    Files of type: Drawing (*.dwg)

    Browse to the correct folder for the downloaded files.

    Select *dining group* from the exercise files.
    Click **Open**.
    Select the *dining group.dwg* from the Blocks palette.
    Right click and select **Insert**.

5.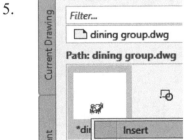

6.

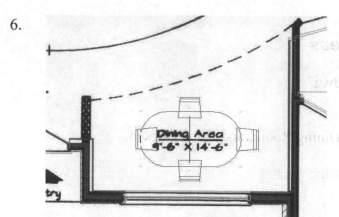

Left click to add the **Dining Set** to the dining room area by the entry.

7.

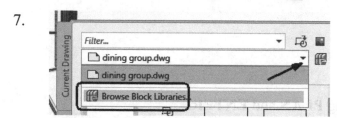

Select the Other Drawing tab.

Select the down arrow and click **Browse Block Libraries.**

8.

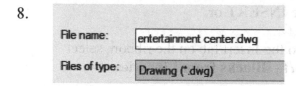

Select *entertainment center* from the exercise files.
Click **Open**.

9.

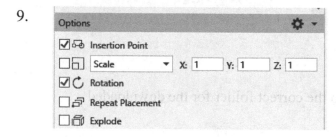

Under Options:
Enable **Insertion Point.**
Enable **Rotation.**

10.

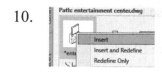

Highlight the preview of the entertainment center in the upper left of the top pane.
Right click and select **Insert**.

11.
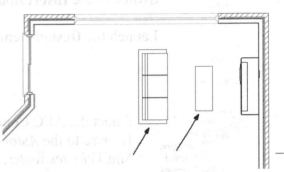
Place in the family room.

12. Add a sofa and coffee table to the
family room using the Design Center.

Save the file as *ex4-2.dwg*.

T²ips Tricks

If the drawing preview in the Design Center is displayed as 2D, then it is a 2D drawing.

Exercise 4-3:

Adding to the Service Areas

Drawing Name: service areas.dwg
Estimated Time: 30 minutes

1. Open *service areas.dwg.*
   Select the Model tab.
   Switch to a top view.

2. Switch to the **Insert** ribbon.

   Launch the **Design Center**.

3. Select the AEC Content tab.
   Browse to the *AutoCAD Architecture/Imperial/Design/Site/Basic Site/Vehicles* folder.

4. Sub-Compact

   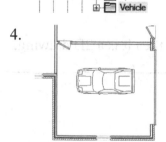

   Drag and drop the *Sub-Compact* car into the garage.

All that remains are the kitchen, bathroom, and laundry areas.

5.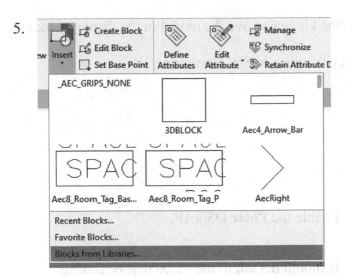

Activate the Insert ribbon.

Select **Insert→Blocks from Libraries**.

6.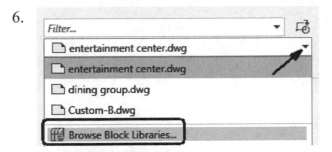

Select the drop-down arrow and **Browse Block Libraries.**

7.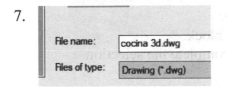

Locate the *cocina 3d.dwg* in the exercise files. Click **Open**.

Cocina means kitchen in Spanish.

8.

In the lower pane:
Enable **Insertion Point.**
Set the **Rotation** to **45.**

9.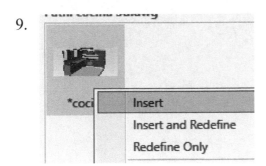

Right click on the drawing in the upper pane.

Select **Insert**.

10.

Place the kitchen above the model in the display window.

Change the view to SW Isometric.

11.

✓ 45.00, 90.00, 135.00, 180.00…
30.00, 60.00, 90.00, 120.00…
22.50, 45.00, 67.50, 90.00…
18.00, 36.00, 54.00, 72.00…
15.00, 30.00, 45.00, 60.00…
10.00, 20.00, 30.00, 40.00…
5.00, 10.00, 15.00, 20.00…

45.00, 90.00, 135.00, 180.00…
135.00, 270.00, 405.00, 540.00…
225.00, 450.00, 675.00, 900.00…
315.00, 630.00, 945.00, 1260.00…

Tracking Settings…

Enable the **Polar** OSNAP.

It should default to the 45.00 degree setting.

12.

Type **SC** for scale.
Select the kitchen block.
Click ENTER to complete the selection.

When prompted for the base point, select the insertion point indicated.

13.

Enter
Cancel
Copy
Reference
Pan
Zoom

Right click and select **Reference**.

14.

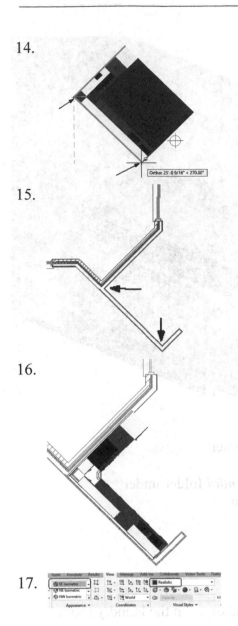

Select the basepoint and the lower right corner as the reference length.

15.

Right click and select Points when prompted for the new length.

Select the corner points indicated for the new length.

16.

Use the MOVE tool to move the kitchen block into the kitchen area.

17.

Switch to a **SE isometric** 3D view.
Change the visual display to **Realistic.**

18. Inspect the kitchen.

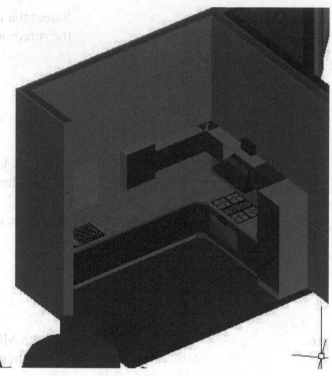

19.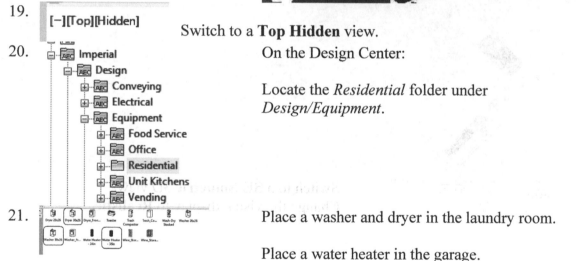

Switch to a **Top Hidden** view.

20. On the Design Center:

Locate the *Residential* folder under *Design/Equipment*.

21. Place a washer and dryer in the laundry room.

Place a water heater in the garage.

22. Add furniture and decorations from the Design Center. Additional blocks are included in the downloaded files.

    You can also search the internet for 3D models you can use to fill in your floor plan.

    The completed floor plan should look similar to this.

    Your floor plan may look differently depending on which furnishings you used.

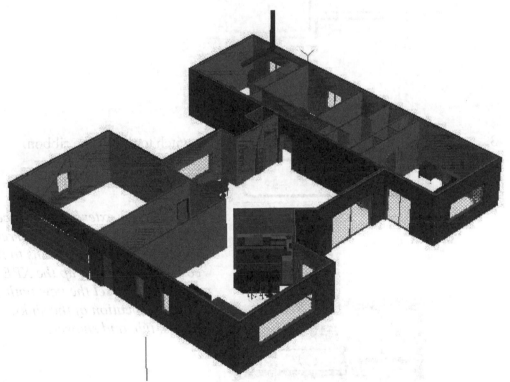

23. Save the file as *ex 4-3.dwg*.

Exercise 4-4:

Adding Components to the Bathrooms

Drawing Name: bathrooms.dwg
Estimated Time: 30 minutes

1. 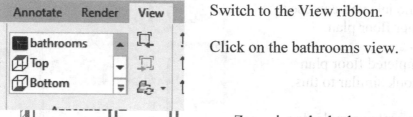 Switch to the View ribbon.

   Click on the bathrooms view.

2. Zoom into the bathroom area.

3. Switch to the Home ribbon.

   Thaw the image layer.

   This file uses external references. If any of the links are not correct, you will need to point the paths to the correct files. Bring up the XREF Manager and set the new paths. Note the location of the sinks, toilets, tub, and shower.

4.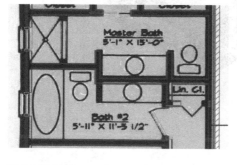

5. Switch to the Insert ribbon.

   Launch the **Design Center**.

6. 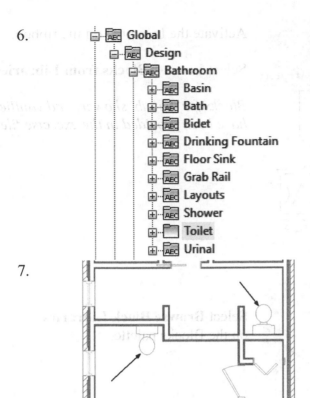 Select the AEC Content tab on the Design Center palette.
Browse to the *AutoCAD Architecture/Global//Design/Bathroom/Toilet* folder.

7. Place a toilet in the two bathrooms.

Highlight the **3D Toilet – Flush Valve**.

Right click and select **Insert**.

8. Left click to place the toilet.
Use the cursor to rotate the toilet into the correct orientation.
Left click to complete placement.

Move the cursor to the second toilet location.
Left click to place the toilet.
Use the cursor to rotate the toilet into the correct orientation.
Left click to complete placement.
Click **ESC** to exit the Insert command.

9. Freeze the image layer.

This will make it easier to place the remaining items.

10. Enable ORTHO, OBJECT TRACKING, and OSNAPS to place the blocks.

11. 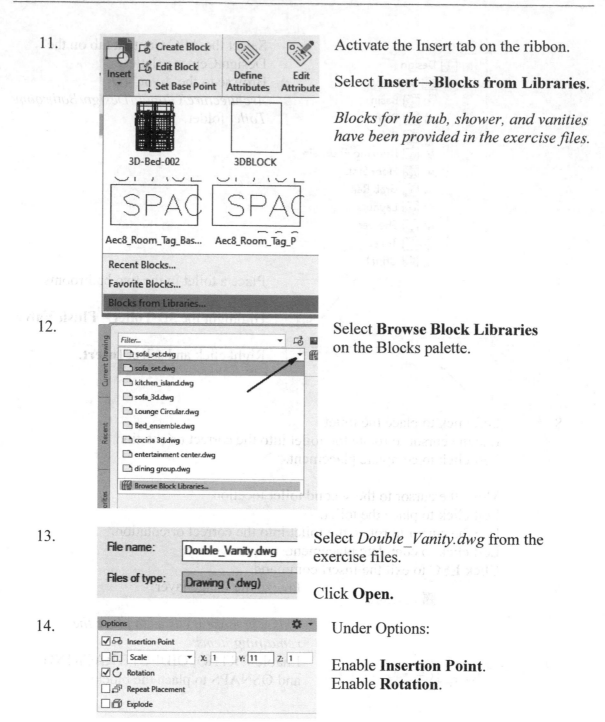 Activate the Insert tab on the ribbon.

Select **Insert→Blocks from Libraries**.

Blocks for the tub, shower, and vanities have been provided in the exercise files.

12. Select **Browse Block Libraries** on the Blocks palette.

13. Select *Double_Vanity.dwg* from the exercise files.

Click **Open.**

14. Under Options:

Enable **Insertion Point**.
Enable **Rotation**.

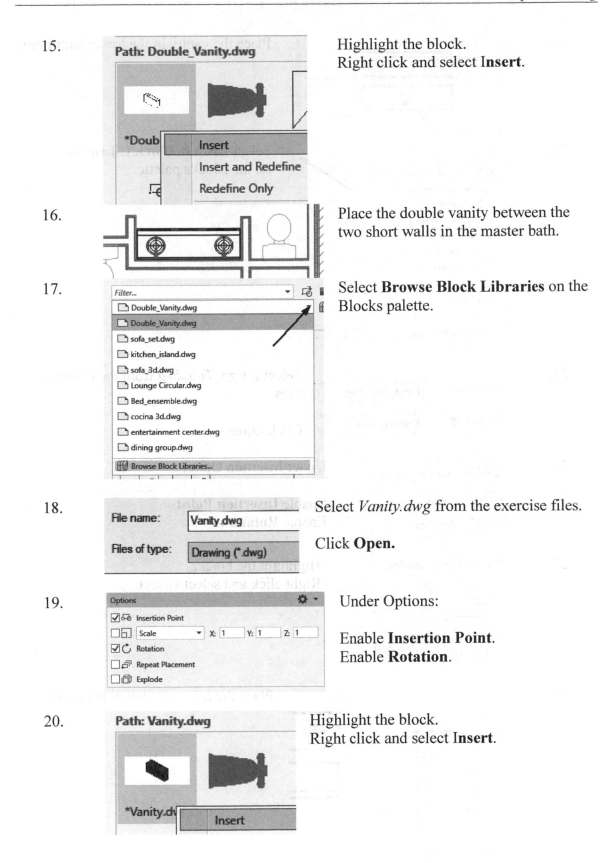

15. Highlight the block.
Right click and select **Insert**.

16. Place the double vanity between the two short walls in the master bath.

17. Select **Browse Block Libraries** on the Blocks palette.

18. Select *Vanity.dwg* from the exercise files.

Click **Open.**

19. Under Options:

Enable **Insertion Point**.
Enable **Rotation**.

20. Highlight the block.
Right click and select **Insert**.

21.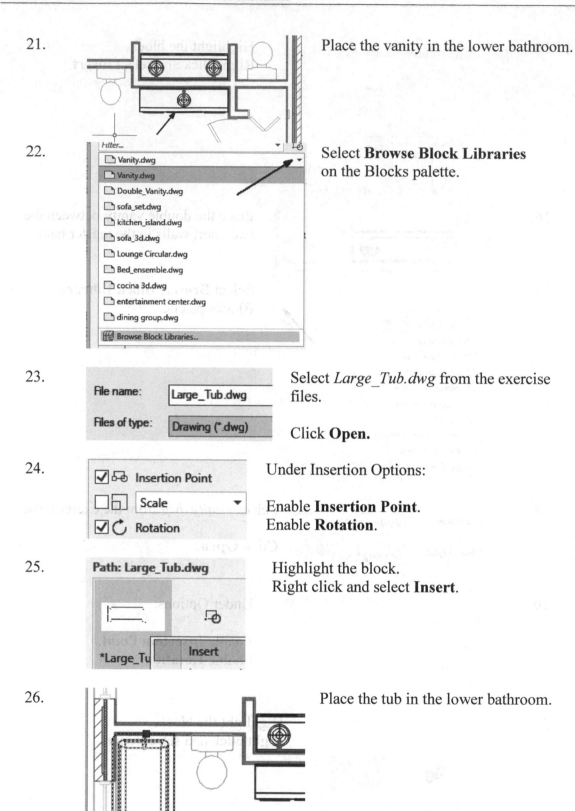

Place the vanity in the lower bathroom.

22.

Select **Browse Block Libraries** on the Blocks palette.

23.

Select *Large_Tub.dwg* from the exercise files.

Click **Open.**

24.

Under Insertion Options:

Enable **Insertion Point**.
Enable **Rotation**.

25.

Highlight the block.
Right click and select **Insert**.

26.

Place the tub in the lower bathroom.

27.

Select **Browse Block Libraries** on the Blocks palette.

28.

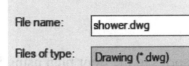

Select *shower.dwg* from the exercise files.

Click **Open**.

29.

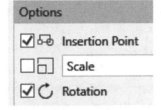

Under Insertion Options:

Enable **Insertion Point**.
Enable **Rotation**.

30.

Highlight the block.
Right click and select **Insert**.

31.

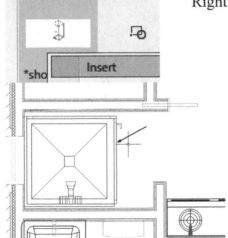

Place the shower in the master bathroom.

Can you use the SCALE/REFERENCE tool to adjust the size of the shower?

32. 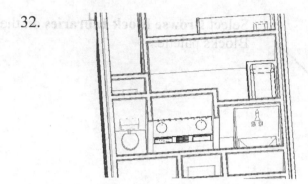 Save as *ex4-4.dwg*.

Extra:

Add towel bars, plants, mirrors, and other accessories to the floor plan using blocks from the design center or downloaded from the internet.

QUIZ 4

True or False

1. Custom content can be located in any subdirectory and still function properly.

2. The sole purpose of the Space Planning process is to arrange furniture in a floor plan.

3. The Design Center only has 3D objects stored in the Content area because ADT is strictly a 3D software.

4. Appliances are automatically placed on the APPLIANCE layer.

5. When you place a wall cabinet, it is automatically placed at the specified height.

6. You can create tools on a tool palette by dragging and dropping the objects from the Design Center onto the palette.

Multiple Choice

7. A residential structure is divided into:

   A. Four basic areas
   B. Three basic areas
   C. Two basic areas
   D. One basic area

8. Kitchen cabinets are located in the _____ subfolder.

   A. Casework
   B. Cabinets
   C. Bookcases
   D. Furniture

9. Select the area type that is NOT part of a private residence:

   A. Bedrooms
   B. Common Areas
   C. Service Areas
   D. Public Areas

10. To set the layer properties of a tool on a tool palette:

   A. Use the Layer Manager
   B. Select the tool, right click and select Properties
   C. Launch the Properties dialog
   D. All of the above

11. Vehicles placed from the Design Center are automatically placed on this layer:

    A. A-Site-Vhcl
    B. A-Vhcl
    C. C-Site Vhcl
    D. None of the above

12. The **Roof** tool is located on this tool palette:

    A. DESIGN
    B. GENERAL DRAFTING
    C. MASSING
    D. TOOLS

ANSWERS:

1) T; 2) F; 3) F; 4) F; 5) T; 6) T; 7) B; 8) A; 9) D; 10) B; 11) C; 12) A

Lesson 5:
Roofs

Roofs can be created with single or double slopes, with or without gable ends, and with or without overhangs. Once you input all your roof settings, you simply pick the points around the perimeter of the building to define your roof outline. If you make an error, you can easily modify or redefine your roof.

You need to pick three points before the roof will begin to preview in your graphics window. There is no limit to the number of points to select to define the perimeter.

To create a gable roof, uncheck the gable box in the Roof dialog. Pick the two end points for the sloped portion of the roof. Turn on the Gable box. Pick the end point for the gable side. Turn off the Gable box. Pick the end point for the sloped side. Turn the Gable box on. Pick the viewport and then Click ENTER. A Gable cannot be defined with more than three consecutive edges.

Roofs can be created using two methods: ROOFADD places a roof based on points selected or ROOFCONVERT which converts a closed polyline or closed walls to develop a roof.

> ➤ If you opt to use ROOFCONVERT and use existing closed walls, be sure that the walls are intersecting properly. If your walls are not properly cleaned up with each other, the roof conversion is unpredictable.
> ➤ The Plate Height of a roof should be set equal to the Wall Height.
> ➤ You can create a gable on a roof by gripping any ridgeline point and stretching it past the roof edge. You cannot make a gable into a hip using grips.

Shape – Select the Shape option on the command line by typing 'S.'	Single Slope – Extends a roof plane at an angle from the Plate Height.	 end elevation view
	Double Slope – Includes a single slope and adds another slope, which begins at the intersection of the first slope and the height specified for the first slope.	 end elevation view
Gable – Select the Gable option on the command line by typing 'G.'	If this is enabled, turns off the slope of the roof place. To create a gable edge, select Gable prior to identifying the first corner of the gable end. Turn off gable to continue to create the roof.	 gable roof end
Plate Height – Set the Plate Height on the command line by typing 'PH.'	Specify the top plate from which the roof plane is projected. The height is relative to the XY plane with a Z coordinate of 0.	
Rise – Set the Rise on the command line by typing 'PR.'	Sets the angle of the roof based on a run value of 12.	A rise value of 5 creates a 5/12 roof, which forms a slope angle of 22.62 degrees.
Slope – Set the Slope on the command line by typing 'PS.'	Angle of the roof rise from the horizontal.	If slope angles are entered, then the rise will automatically be calculated.

Upper Height – Set the Upper Height on the command line by typing 'UH.'	This is only available if a Double Slope roof is being created. This is the height where the second slope will start.	
Rise (upper) – Set the Upper Rise on the command line by typing 'UR.'	This is only available if a Double Slope roof is being created. This is the slope angle for the second slope.	A rise value of 5 creates a 5/12 roof, which forms a slope angle of 22.62 degrees.
Slope (upper) – Set the Upper Slope on the command line by typing 'US.'	This is only available if a Double Slope roof is being created. Defines the slope angle for the second slope.	If an upper rise value is set, this is automatically calculated.
Overhang – To enable on the command line, type 'O.' To set the value of the Overhang, type 'V.'	If enabled, extends the roofline down from the plate height by the value set.	

	The Floating Viewer opens a viewer window displaying a preview of the roof.
	The match button allows you to select an existing roof to match its properties.
	The properties button opens the Roof Properties dialog.
	The Undo button allows you to undo the last roof operation. You can step back as many operations as you like up to the start.
	Opens the Roof Help file.

Exercise 5-1:

Creating a Roof using Existing Walls

Drawing Name: roofs1.dwg
Estimated Time: 10 minutes

This exercise reinforces the following skills:

- ❑ Roof
- ❑ Roof Properties
- ❑ Visual Styles

1.   Open *roofs1.dwg.*

2.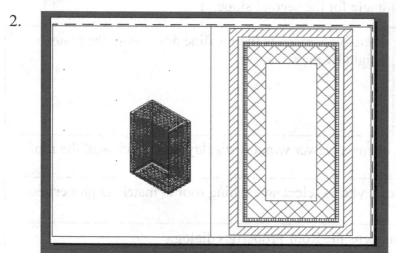

    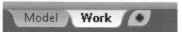

    Select the Work icon. Activate the right viewport that shows the top or plan view.

3.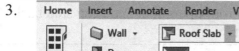

    Select the **Roof** tool from the Home ribbon.

4.

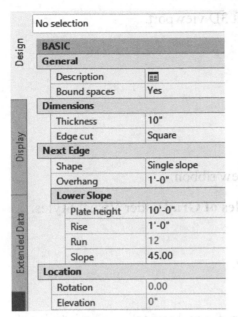

Expand the Dimensions section.
Set the Thickness to **10″ [254.00 mm]**.
Set the Shape to **Single slope**.
Set the Overhang to **1′-0″ [609.6 mm]**.

Dimensions	
Thickness	254.00
Edge cut	Square
Shape	Single slope
Overhang	609.60
	☷ Edges/Faces

5.

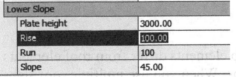

Expand the Lower Slope section.
Set the Plate height to **10′-0″ [3000 mm]**.

The plate height determines the level where the roof rests.

6.

Lower Slope	
Plate height	3000.00
Rise	100.00
Run	100
Slope	45.00

Set the Rise to **1′-0″ [100 mm]**.

7.

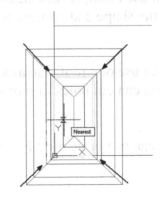

Pick the corners indicated to place the roof.

Click **Enter**.

8.  Activate the left 3D viewport.

9.  Activate the View ribbon.

Select the **Shades of Gray** under Visual Styles.

10. Save as *ex5-1.dwg*.

In order to edit a roof, you have to convert it to Roof Slabs.

Roof Slabs

A roof slab is a single face of a roof. Roof slab elements differ from roof elements in that each roof slab is independent of other roof slabs. When you use multiple roof slabs to model a roof, you have more flexibility as you can control the shape and appearance of each slab or section of the roof.

While roof slabs are independent, they can interact. You can use one roof slab as a trim or extend boundary. You can miter roof slabs together. You can cut holes in roof slabs and add or subtract mass elements.

You can use grips to modify the shapes of roof slabs. You can use basic modify tools, like move or rotate, to reposition roof slabs.

Exercise 5-2:

Roof Slabs

Drawing Name: roof_slabs.dwg
Estimated Time: 15 minutes

This exercise reinforces the following skills:

- ❏ Convert to Roof
- ❏ Roof Slab Tools

1.   Open *roof_slabs.dwg*.

2.   Select the **Roof** tool from the Home ribbon.

3.

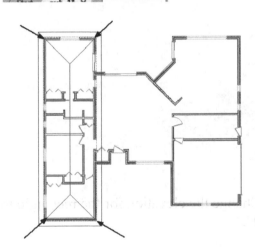

Select the four corners on the left wing to place a roof.

Click ENTER to complete the command.

4.

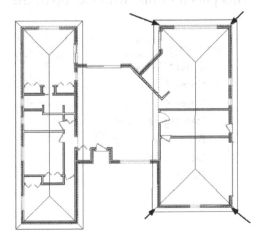

Repeat ROOFADD

Recent Input

Repeat and select the four corners on the right wing to place a roof.

5.

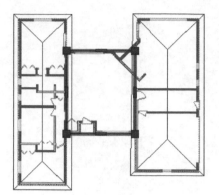

Draw a rectangle around the middle section of the floor plan using the rectangle tool.

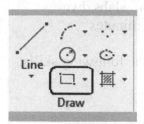

6.

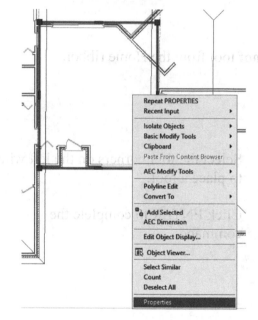

Select the rectangle.
Right click and select **Properties.**

7.

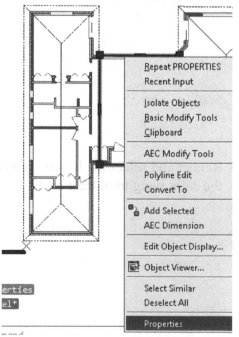

Change the elevation for the rectangle to 10'-0".

This puts it in-line with the top of the walls.

Geometry	
Material	ByLayer
Geometry	
Current Vertex	1
Vertex X	27'-8 3/8"
Vertex Y	51'-9 5/8"
Start segment width	0"
End segment width	0"
Global width	0"
Elevation	10'-0"

8. 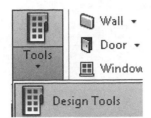 Launch the Design Tools from the Home ribbon.

9. 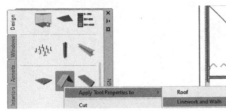 Locate the **Roof** tool on the Design palette by scrolling down.

Right click and select **Apply Tool Properties to→Linework and Walls.**

10. 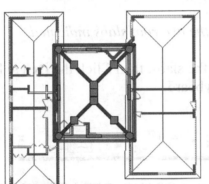 Select the rectangle.
Click ENTER to complete the selection.
When prompted to erase the linework, select **Yes**.
Click ENTER.
The middle roof is placed.

Click ESC to release any selection.

11.  Select all three roofs.

The roofs will highlight.

12.  Select **Convert** from the Roof ribbon.

13. 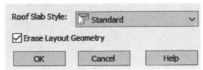 Enable the **Erase layout geometry** checkbox.

Click **OK**.

14. The roofs will change appearance slightly. They now consist of individual slabs.

15. Save as *ex5-2.dwg*.

Exercise 5-3:

Modify Roof Line

Drawing Name: modify_roof.dwg
Estimated Time: 15 minutes

This exercise reinforces the following skills:

- ❑ Modify Roof Slabs
- ❑ Modify Roof Line

1.   Open *modify_roof.dwg.*

2.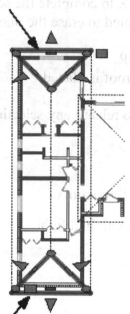

Erase the two roof slabs indicated.

Select the slabs, then Click the **Delete** key on the keyboard.

Verify that ORTHO, OTRACK and OSNAP are enabled on the status bar.

3.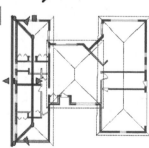

Select the left roof slab.

4.

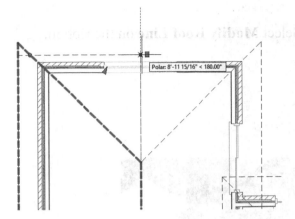

Drag the point/grip at the bottom of the triangle up until it is horizontal to the upper point.

Look for the input that displays the angle at 180°.

Click ESC to release the selection.

5.

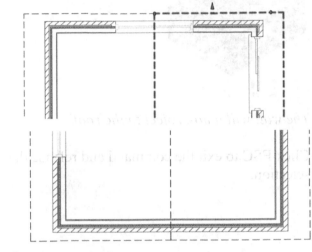

Repeat for the right side.

6.

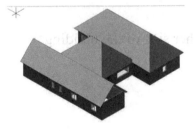

Repeat for the bottom side.

If you are having problems creating a straight edge, draw a horizontal line and snap to the line. Then, erase the line.

7.

Switch to a 3D view to see how the roof has changed.

[−][SW Isometric][Current]

View Style shown is Hidden, Materials On, Realistic Face Style.

Hint: You can save these settings as a Visual Style.

8.

Select the wall indicated.

9.   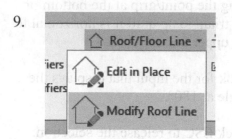   Select **Modify Roof Line** on the ribbon.

10. Type **A** for Autoproject.
     Select the two roof slabs.
     Click ENTER.

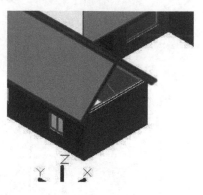

11.    *The wall will auto-project to the roof.*

     Click ESC to exit the command and release the
     selection.

12.   Rotate the model so you can see the other side of the building.

13. 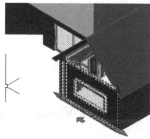   Select the wall indicated.

14. 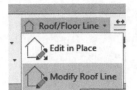 Select **Modify Roof Line** from the ribbon.

15. Type **A** for Autoproject. Select the two roof slabs. Click ENTER.

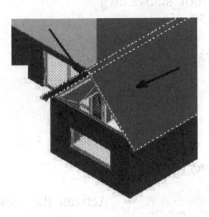

16. 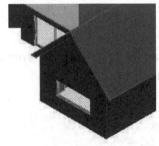 The wall will auto-project to the roof.

Click ESC to exit the command.

Release the selection.

17. Save as *ex5-3.dwg*.

Exercise 5-4:

Trimming Roof Slabs

Drawing Name: roof_slabs2.dwg
Estimated Time: 30 minutes

This exercise reinforces the following skills:

-   Roof Slabs
-   Trim
-   Grips

1.   📂   Open *roof_slabs2.dwg*.

2.   Activate the **View** ribbon.

Select the **Top** view.

You can also use the control at the top left corner of the display window to change the view orientation.

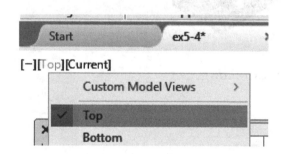

3.

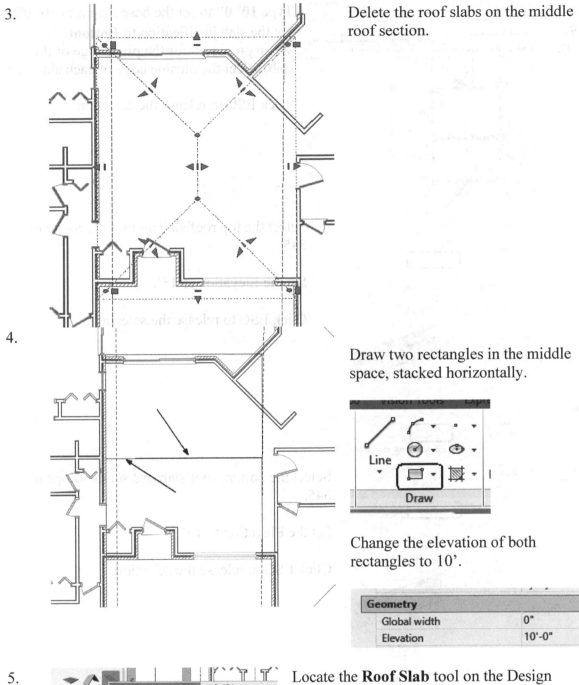

Delete the roof slabs on the middle roof section.

4.

Draw two rectangles in the middle space, stacked horizontally.

Change the elevation of both rectangles to 10'.

Geometry	
Global width	0"
Elevation	10'-0"

5.

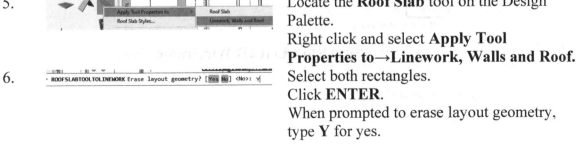

Locate the **Roof Slab** tool on the Design Palette.
Right click and select **Apply Tool Properties to→Linework, Walls and Roof.**

6.

ROOFSLABTOOLTOLINEWORK Erase layout geometry? [Yes No] <No>: Y

Select both rectangles.
Click **ENTER**.
When prompted to erase layout geometry, type **Y** for yes.

7.

ROOFSLABTOOLTOLINEWORK Creation mode [Direct Projected] <Projected>:

Click **ENTER** to accept Projected.

8.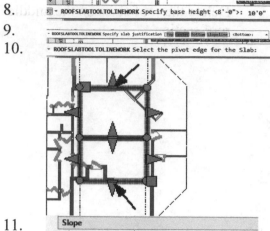
    Type **10' 0"** to set the base height to 10' 0".

9. Set the slab justification to **Bottom**.

10. When prompted for the pivot edge of the slab, select the outside edge of each slab.

    Click ESC to release the selection.

11.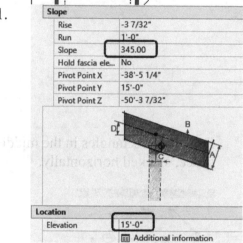
    Select the top roof slab and set the Slope to **345**.

    Set the Elevation to **15'**.

    Click ESC to release the selection.

12.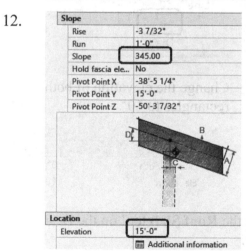
    Select the bottom roof slab and set the Slope to **345**.

    Set the Elevation to **15'**.

    Click ESC to release the selection.

13. Switch to a **Left 2D Wireframe** view.

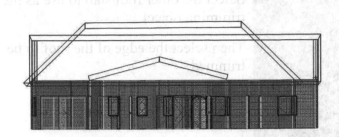

You can see how the slabs are positioned.

Adjust their position if necessary.

14. [−][SW Isometric][Hidden]   Switch to a SW Isometric Hidden view.

15. 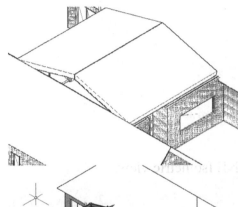   The slabs look OK.

If they don't look proper, you may need to trim them.

16. 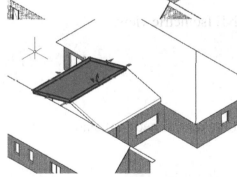   Select a roof slab.

17. 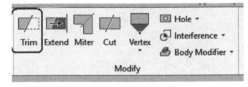   Select **Trim** on the ribbon.

18.

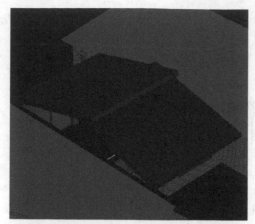

Select the other roof slab to use as the trimming object.

Then select the edge of the roof to be trimmed.

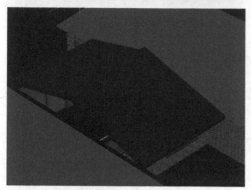

The roof slab adjusts.

19.

Activate the NE Isometric view.

20.

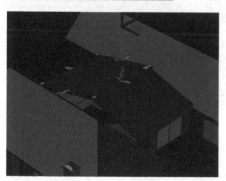

Select the untrimmed roof slab.

21.

Select **Trim** on the ribbon.

22.

Select the other roof slab to use as the trimming object.

Then select the edge of the roof to be trimmed.

The roof slab adjusts.

23.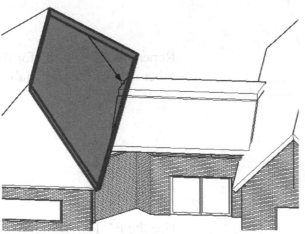

Select one of the middle roof slabs.

Select **Extend** from the ribbon.

When prompted for object to extend to: select one of the adjoining roof slabs.

24.

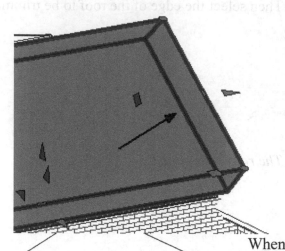

When prompted to select an edge
to lengthen, select the upper edge
of the roof slab closest to the
adjoining roof slab used as the
boundary.

25.

When prompted to select a second edge,
select the lower edge of the roof slab to be
extended.

26.

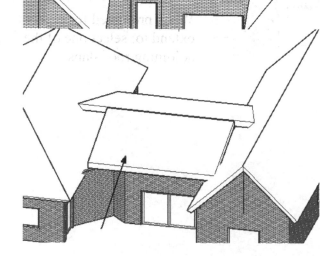

Repeat the EXTEND for the
other side of the roof slab.

27.

Use the EXTEND tool to clean
up the roof slab on the other side.

*If you struggle getting one side of
the roof correct, delete that roof
slab. Switch to a TOP view and
mirror the roof slab that looks
correct.*

28.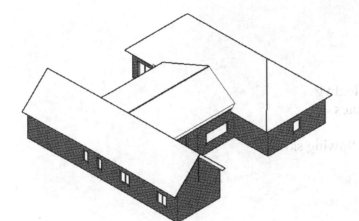

Save as *ex5-4.dwg*.

Exercise 5-5:

Add a Hole to a Roof

Drawing Name: add_hole.dwg
Estimated Time: 10 minutes

This exercise reinforces the following skills:

- ❑ Roof Slabs
- ❑ Add Hole

1. 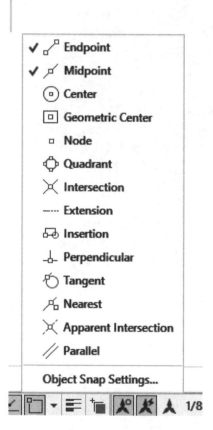   Open *add_hole.dwg*.

2. [−][Top][Hidden]   Switch to a **Top** view.

3.

Verify that only Endpoint and Midpoint should be enabled on the OSNAP settings.

4.

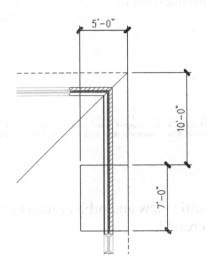

Draw a rectangle **10' 0"** below the top edge of the far right roof slab.

The rectangle should be 5' 0" wide and 7' 0" high.

Do not add dimensions...those are for reference only.

If you use lines to place, use the PE polyline edit command to create a closed polyline.

5.

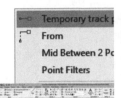

Turn ORTHO **ON**.
Start the Rectangle command.
Right click and select TEMPORARY TRACK POINT.

6.

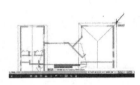

Left pick to select the upper right corner of the roof. *Look for the small green X to indicate tracking is enabled.*

Drag the mouse straight down and type 10' to start the rectangle 10' below the tracking point.

7.

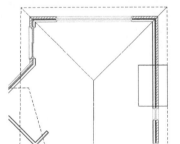

Type **@-5',-7'** to complete the rectangle.

This means you are moving five feet to the left and seven feet down to complete the rectangle.

8.

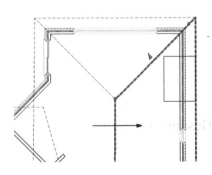

Select the far right roof slab.

9. 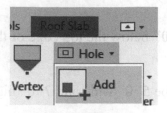   Select **Hole→Add** from the ribbon.

10. Select the rectangle.
    Click **ENTER**.
    Type **Y** to erase the layout
    geometry.

⌐ ROOFSLABADDHOLE Select closed polylines or connected solid objects to define holes:

11. 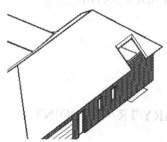   Switch to an isometric view and orbit around to see
    the hole that was created.

12. Save as *ex5-5.dwg*.

Exercise 5-6:

Add a Gabled Roof

Drawing Name: gable.dwg
Estimated Time: 20 minutes

This exercise reinforces the following skills:

- ❑ Roof
- ❑ Modify Wall

1.   Open *gable.dwg*.

2. `[−][Top][Hidden]`   Switch to a **Top** view.

3.   Select the **Roof** tool from the Design palette.
     Roof

4.

Edge cut	Square
Next Edge	
Shape	Single slope
Overhang	1'-0"

We change the style of roof for each side to create a gable roof.

Set the Shape to **Single slope**.

5.

Select Points **A** and **D**.

6.

Edge cut	Square
Next Edge	
Shape	Single slope
Overhang	1'-0"

Set the Shape to **Single slope**.
Select Points **D** and **C**.

7.

No selection	
BASIC	−
General	−
Description	
Bound spaces	Yes
Dimensions	−
Thickness	10"
Edge cut	Square
Next Edge	−
Shape	Gable
Overhang	1'-0"
Location	−

In the Properties pane,
set the Shape to **Gable**.

8.

Select the two vertex points of the rectangle on the right side.

Select Point B and then Point C.

9.

Edge cut	Square
Next Edge	
Shape	Single slope
Overhang	1'-0"

Set the Shape to **Single slope.**

10.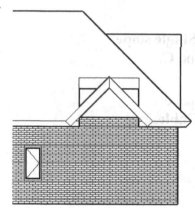

Select Point D and then Point C.

The gable roof is created.

Click **ENTER**.

11. [−][Right][Hidden]   Switch to a Right view.

12.

You see the gable that was created.

Select the gable.

13.

Lower Slope	
Plate height	17'-0"
Rise	1'-0"
Run	12
Slope	45.00

Set the Plate Height to **17' 0"** on the Properties palette.

This will raise the gable.

14.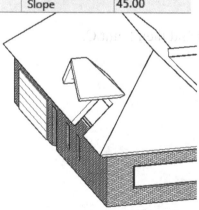

You can orbit the view slightly to check the gable position.

15.

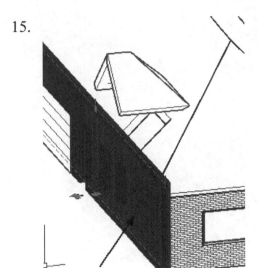

Select the lower wall below the gable.

16.

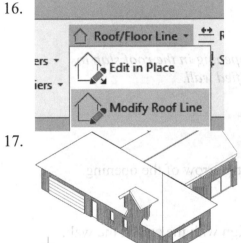

Select **Modify Roof Line** on the ribbon.

17.

Select **Auto project**.

Select the gabled roof. Click ENTER.

The wall will adjust.

18. Save as *ex5-6.dwg*.

Exercise 5-7:

Modify a Hole in a Roof Slab

Drawing Name: hole_modify.dwg
Estimated Time: 10 minutes

This exercise reinforces the following skills:

❑   Roof Slabs

1.   Open hole_modify.dwg.

2.   **[–][Top][2D Wireframe]**   Switch to a **Top 2D Wireframe** view.

3.   *We need to adjust the opening in the roof slab to accommodate the modified wall.*

Select the roof slab.

4.   Select the middle top grip arrow of the opening.

5.   Move the edge up to align with the end of the wall.

6.   Repeat on the bottom edge of the opening.

Select the slab.
Select the middle bottom grip arrow.
Move the edge down to align with the end of the wall.

7. 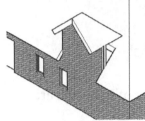 Verify that the opening aligns with the wall ends.

8. [−][NE Isometric][Hidden]

Switch to a NE Isometric Hidden view.

Note how the opening has updated.

9. Save as *ex5-7.dwg*.

Exercise 5-8:

Add Walls to a Gabled Roof

Drawing Name: gable_walls.dwg
Estimated Time: 20 minutes

This exercise reinforces the following skills:

❑ Walls
❑ Wall Properties
❑ Global Cut Plane Display

1. 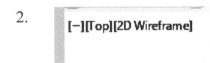 Open *gable_walls.dwg*.

2. [−][Top][2D Wireframe]

Switch to a **Top 2D Wireframe** view.

3. 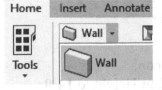 Select the **Wall** tool from the Home ribbon.

4.

	Browse...
Style	🗋 Stud-4 Rigid-1.5 Air-1 Brick-4
Bound spaces	By style (Yes)
Cleanup automatically	Yes

In the Properties pane:
Set the Style to **Stud-4 Rigid-1.5 Air- 1 Brick 4.**

5.

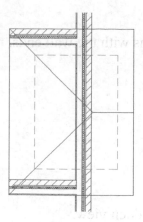

Place two short walls to act as the sides of the gable.

The top wall should be left justified.
The bottom wall should be right justified.

Dimensions		
A	Width	11 1/8"
B	Base height	10'-0"
C	Length	2'-11 3/8"
	Justify	☐ Right

Verify that the exterior sides of the walls are located correctly.

6.

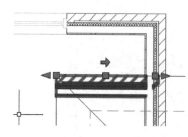

Select the two short walls.

Change their elevations to **10'** using the Properties pane.

Location	
Rotation	0.00
Elevation	10'-0"
	🔲 Additional information

When the elevation changes, they will disappear from the view.

7.

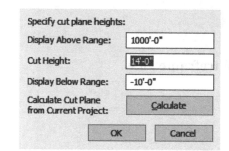

Specify cut plane heights:

Display Above Range:	1000'-0"
Cut Height:	14'-0"
Display Below Range:	-10'-0"
Calculate Cut Plane from Current Project:	Calculate

OK   Cancel

To see the walls in the Plan view, you need to change the global cut plane.

Medium Detail ▼   14'-0"

On the Status bar:
Select the Cut Plane icon.
In the dialog, set the Cut Height to **14'0**.
Click **OK**.

8.

Position the short walls so they are *completely* inside the gabled roof boundaries. Otherwise, we can't use the roof to trim up the walls.

Use FILLET to clean up the corners between the short walls and the exterior wall.

9.

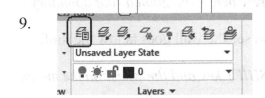

Select the **Layer Properties** dialog from the Home ribbon.

10.

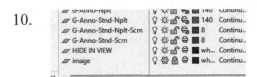

Create a layer called **HIDE IN VIEW**.

We will use this layer to temporarily hide elements to allow easier editing.

11.

-][NW Isometric][Hidden]

Switch to a **NW Isometric Hidden** view.

12.

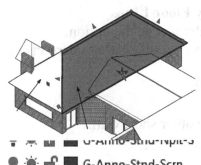

Select the north roof slab next to the gable and the far side roof slab and assign to the **HIDE IN VIEW** layer.

13.

Set the **HIDE IN VIEW** layer to frozen.

14.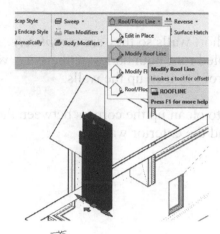

Select a short wall.
From the ribbon:
Select **Modify Roof Line**.
Select **Auto-Project** as the option.
Select the gabled roof.

Click **ENTER**.

Repeat for the other short wall.

If the walls don't trim properly check to see that they are entirely below the gabled roof boundary.

15.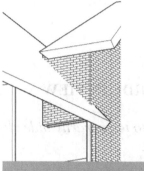

If you orbit the model you see the wall goes below the roof.

To orbit, hold down the SHIFT key and the wheel on the mouse at the same time.

16.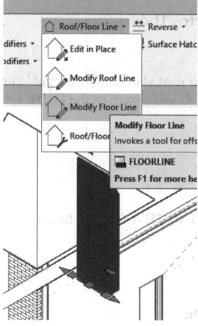

Select one of the short walls.

From the ribbon:
Select **Modify Floor Line**.
Select **Auto-Project** as the option.
Select the sloped roof slab that intersects the wall.

Click **ENTER**.

Repeat for the other short wall.

17.

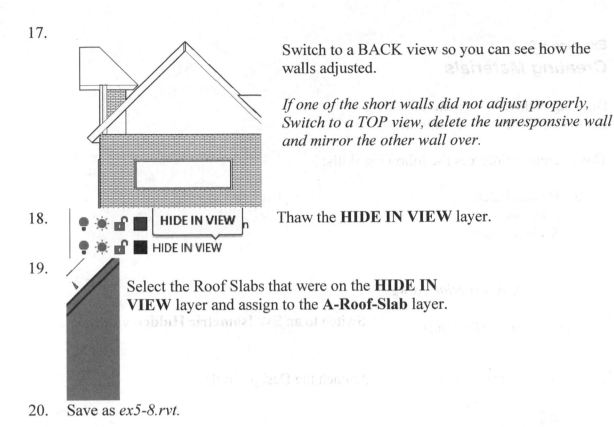

Switch to a BACK view so you can see how the walls adjusted.

If one of the short walls did not adjust properly, Switch to a TOP view, delete the unresponsive wall and mirror the other wall over.

18. Thaw the **HIDE IN VIEW** layer.

19. Select the Roof Slabs that were on the **HIDE IN VIEW** layer and assign to the **A-Roof-Slab** layer.

20. Save as *ex5-8.rvt*.

Rendering Materials in ACA

Materials represent substances such as steel, concrete, cotton and glass. They can be applied to a 3D model to give the objects a realistic appearance. Materials are useful for illustrating plans, sections, elevations and renderings in the design process. Materials also provide a way to manage the display properties of object styles.

The use of materials allows you to more realistically display objects. You need to define the display of a material such as brick or glass only once in the drawing or the drawing template and then assign it to the component of an object where you want the material to display. You typically assign materials to components in the style of an object such as the brick in a wall style. Then whenever you add a wall of that style to your drawing, the brick of that wall displays consistently. Defining materials in an object style can provide control for the display of objects across the whole project. When the characteristics of a material change, you change them just once in the material definition and all objects that use that material are updated. With the Material tool you can apply a material to a single instance of the object.

You can take advantage of Visual Styles, Rendering Materials, Lights, and Cameras in AutoCAD® Architecture. Materials provide the ability to assign surface hatches to objects. Surface hatches can be displayed in model, elevation, and section views. This is helpful to clearly illustrate sections and elevations.

Exercise 5-9:
Creating Materials

Drawing Name: materials1.dwg
Estimated Time: 20 minutes

This exercise reinforces the following skills:

    ❑ Design Palette
    ❑ Materials
    ❑ Style Manager

1.   Open *materials1.dwg*.

2.   `[–][SW Isometric][Hidden]`   Switch to an **SW Isometric Hidden** view.

3.   Launch the Design Tools.

4.   Right click on the title bar of the Design Tools and enable **All Palettes**.

5.    Right click on the tabs of the Design Tools and enable **Materials**.

6.

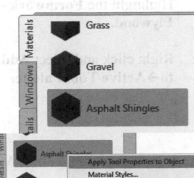

Locate the **Asphalt Shingles** material.

7.

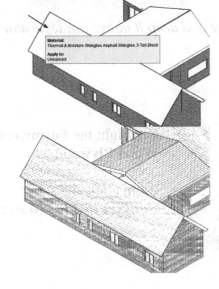

Right click on the **Asphalt Shingles.**

Select **Apply Tool Properties to Object.**

8.

Select one of the roof slabs.

Click **ENTER.**

[−][SW Isometric][Realistic]

Change the display to **Realistic**.

The roof slab updates with the new material definition.

The material definition has now been added to the drawing so it can be used in the Style Manager.

9.

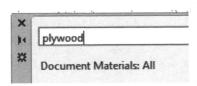

Type **MATERIALS**.

This brings up ACA's Materials Browser.

10.

Type **plywood** in the search field.

11. 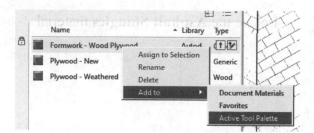 Highlight the **Formwork – Wood Plywood.**

Right click and select **Add to→Active Tool Palette.**

12. 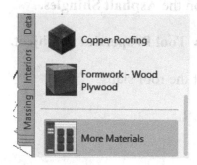 The material appears at the bottom of the Materials palette.

You can drag and drop it above the separator line.

13. 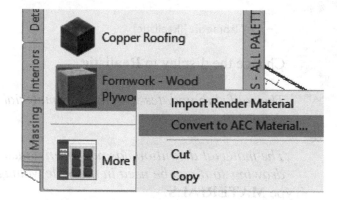 Highlight the **Formwork – Wood Plywood.**

Right click and select **Convert to AEC Material.**

14. Set the AEC Material to use as a Template to **Standard.**

Click **OK**.

15.

Tool Style Update - Save Drawing   ✕

The style for this tool has not been saved with
the drawing.

To ensure that the tool will work properly, the style must be
saved first. You should save this drawing now.

Close

Click **Close**.

Save the drawing file.

16.

rubber

Type **rubber** in the search field.

17.

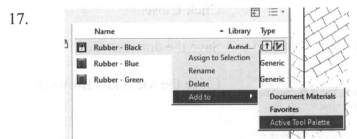

Highlight the **Rubber-Black.**

Right click and select **Add
to→Active Tool Palette.**

18.

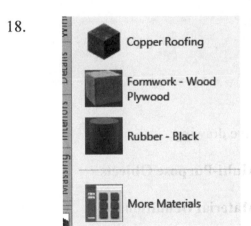

The material appears at the bottom of the Materials
palette.

You can drag and drop it above the separator line.

19.

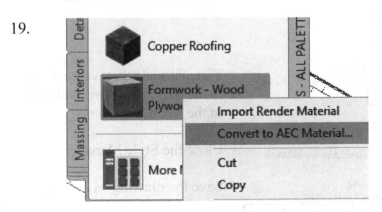

Highlight the **Rubber -
Black.**

Right click and select
Convert to AEC Material.

20.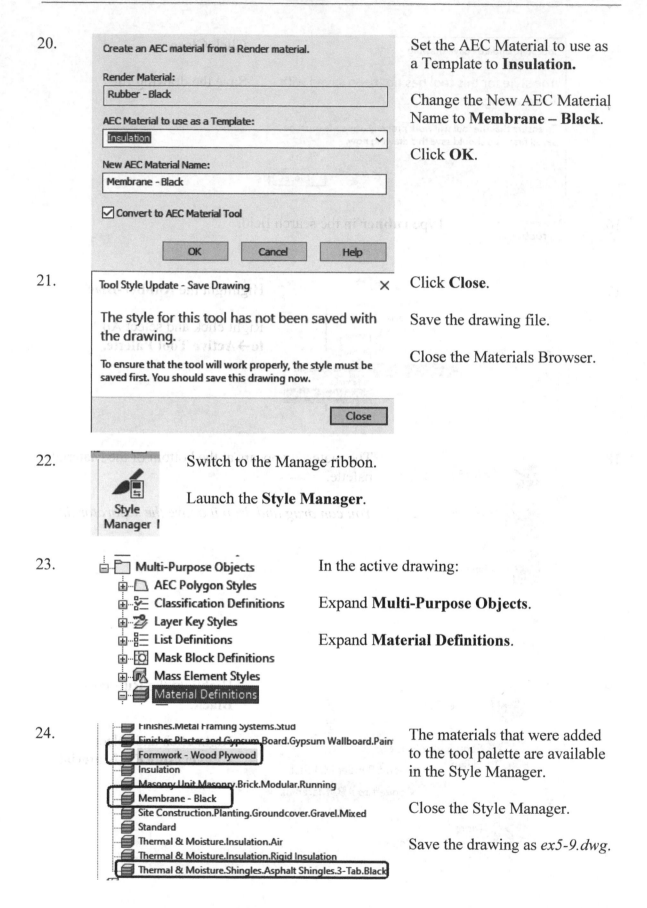

Set the AEC Material to use as a Template to **Insulation.**

Change the New AEC Material Name to **Membrane – Black**.

Click **OK**.

21.

Click **Close**.

Save the drawing file.

Close the Materials Browser.

22.

Switch to the Manage ribbon.

Launch the **Style Manager**.

23.

In the active drawing:

Expand **Multi-Purpose Objects**.

Expand **Material Definitions**.

24.

The materials that were added to the tool palette are available in the Style Manager.

Close the Style Manager.

Save the drawing as *ex5-9.dwg*.

Exercise 5-10:

Roof Slab Styles

Drawing Name: slab_style1.dwg
Estimated Time: 10 minutes

This exercise reinforces the following skills:

☐ Roof Slab Styles
☐ Design Palette
☐ Materials
☐ Style Manager

1.  Open *slab_style1.dwg.*

2. 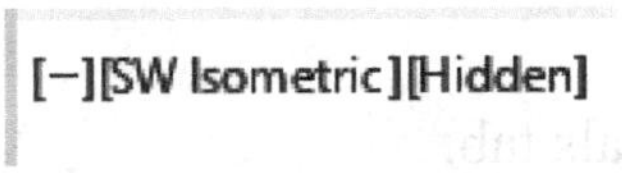 [−][SW Isometric][Hidden]   Switch to an **SW Isometric Hidden** view.

3. 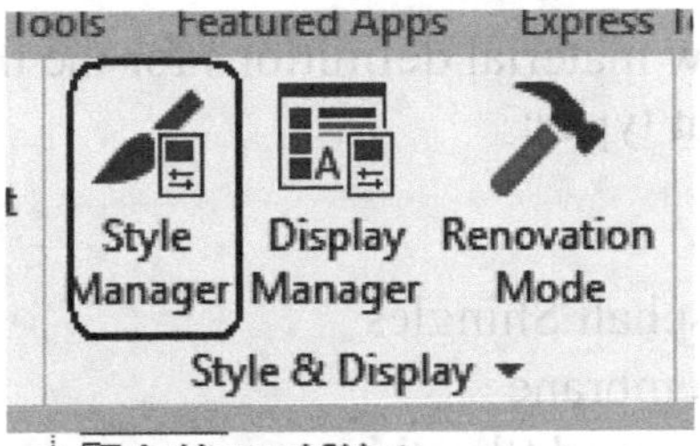 Activate the Manage ribbon.

   Select the **Style Manager**.

4. 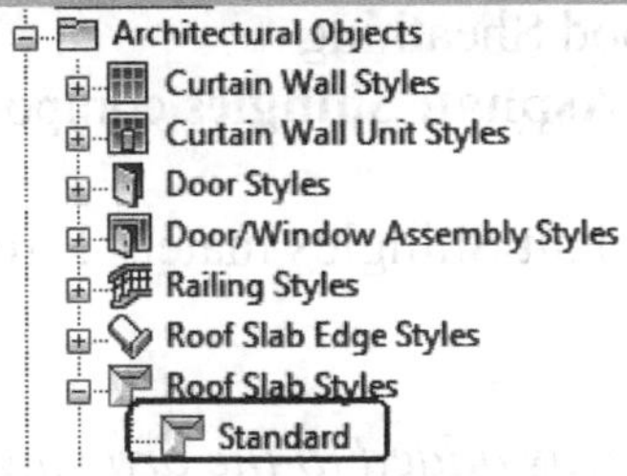 Locate the Roof Slab Styles under Architectural Objects.

   Note that there is only one slab style available in the drawing.

5.  Right click on the Roof Slab Style and select **New**.

6.  Rename the style **Asphalt Shingles**.

7. 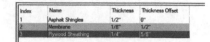   Define three components:
- Asphalt Shingles with a thickness of ½" and Thickness Offset of 0.
- Membrane with a Thickness of 1/8" and Thickness offset of ½".
- Plywood Sheathing with a thickness of ¼" and Thickness offset of 5/8".

8.   *Use the preview to check the placement of each component.*

9.    Select the Materials tab.

Create new material definitions for the three new component types:

- Asphalt Shingles
- Membrane
- Plywood Sheathing

10. 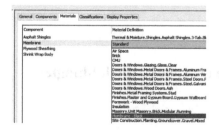   Highlight the **Asphalt Shingles** component.

Select the Asphalt Shingles material from the drop-down list.

This material was added to the drawing when it was applied to the rood slab.

11.   Highlight the **Membrane** component.

Select the Membrane - Black material from the drop-down list.

12.

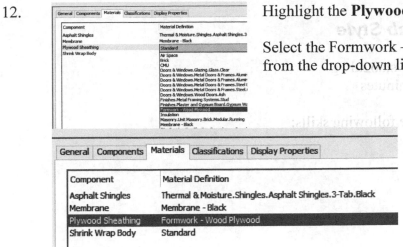

Highlight the **Plywood Sheathing** component.

Select the Formwork – Wood Plywood material from the drop-down list.

The assigned materials should look like this.

Click **OK**.
Close the Style Manager.

13.   Save as *ex5-10.dwg*.

Exercise 5-11:

Applying a Roof Slab Style

Drawing Name: roof_slab2.dwg
Estimated Time: 10 minutes

This exercise reinforces the following skills:

- ❑ Roof Slab Styles
- ❑ Select Similar
- ❑ Properties

1.   Open *roof_slab2.dwg*.

2.   [−][SW Isometric][Hidden]   Switch to an **SW Isometric Hidden** view.

3.
Edit Object Display...
Edit Roof Slab Edge Style...
Edit Roof Slab Style...
Copy Roof Slab Style and Assign...
Object Viewer...
Select Similar
Count Selection
Deselect All

Select a roof slab.
Right click and select **Select Similar**.

4.   *All the roof slabs are selected, but not the gabled roof. That is because it is a roof object and not a roof slab object.*

Right click and select **Properties**.

5.
Search...
Style   Standard
Bound spaces   Asphalt Shingles
Shadow display   Standard

On the Properties palette,
set the Style to **Asphalt Shingles**.

Click **ESC** to release the selection.

6.

Select the gabled roof.

Select **Convert** on the ribbon to convert to roof slabs.

7.

Select **Asphalt Shingles** from the drop-down list.
Enable **Erase Layout Geometry**.
Click **OK**.
Click **ESC** to release the selection.

8.

Save as *ex5-11.dwg*.

Notes:

QUIZ 5

True or False

1. Roof slabs are a single face of a roof, not an entire roof.

2. The thickness of a roof slab can be modified for an individual element without affecting the roof slab style.

3. If you change the slope of a roof slab, it rotates around the slope pivot point.

4. Once you place a roof or roof slab, it cannot be modified.

5. You can use the TRIM and EXTEND tools to modify roof slabs.

Multiple Choice

6. Materials are displayed when the view setting is: (select all that apply)

    A. wireframe
    B. Shaded
    C. Hidden
    D. Rendered

7. Roof slab styles control the following properties: (select all that apply)

    A. Slab Thickness
    B. Slab Elevation
    C. Physical components (concrete, metal, membrane)
    D. Materials

8. Identify the tool shown.

    A. Add Roof
    B. Add Roof Slab
    C. Add Gable
    D. Modify Roof

9. To change the elevation of a slab:

    A. Modify the slab style.
    B. Use Properties to change the Elevation.
    C. Set layer properties.
    D. All of the above.

10. Roof slabs can be created: (select all that apply)

    A. Independently (from scratch)
    B. From Existing Roofs
    C. From Walls
    D. From Polylines

ANSWERS:

1) T; 2) T; 3) T; 4) F; 5) T; 6) A, B, D; 7) A, C, D; 8) B; 9) B; 10) A, B, C

Lesson 6:
Structural Members

Architectural documentation for residential construction will always include plans (top views) of each floor of a building, showing features and structural information of the floor platform itself. Walls are located on the plan but not shown in structural detail.

Wall sections and details are used to show:

- the elements within walls (exterior siding, sheathing, block, brick or wood studs, insulation, air cavities, interior sheathing, trim)
- how walls relate to floors, ceilings, roofs, and eaves
- openings within the walls (doors/windows with their associated sills and headers)
- how walls relate to openings in floors (stairs).

Stick-framed (stud) walls usually have their framing patterns determined by the carpenters on site. Once window and door openings are located on the plan and stud spacing is specified by the designer (or the local building code), the specific arrangement of vertical members is usually left to the fabricators and not drafted, except where specific structural details require explanation.

The structural members in framed floors that have to hold themselves and/or other walls and floors up are usually drafted as framing plans. Designers must specify the size and spacing of joists or trusses, beams and columns. Plans show the orientation and relation of members, locate openings through the floor and show support information for openings and other specific conditions.

In the next exercises, we shall create a floor framing plan. Since the ground floor of our one-story lesson house has already been defined as a concrete slab, we'll assume that the ground level slopes down at the rear of the house and create a wood deck at the sliding door to the family room. The deck will need a railing for safety.

Autodesk AutoCAD Architecture includes a Structural Member Catalog that allows you to easily access industry-standard structural shapes. To create most standard column, brace, and beam styles, you can access the Structural Member Catalog, select a structural member shape, and create a style that contains the shape that you selected. The shape, similar to an AEC profile, is a 2D cross-section of a structural member. When you create a structural member with a style that you created from the Structural Member Catalog, you define the path to extrude the shape along.

You can create your own structural shapes that you can add to existing structural members or use to create new structural members. The design rules in a structural member style allow you to add these custom shapes to a structural member, as well as create custom structural members from more than one shape.

All the columns, braces, and beams that you create are sub-types of a single Structural Member object type. The styles that you create for columns, braces, and beams have the same Structural Member Styles style type as well. When you change the display or style of a structural member, use the Structural Member object in the Display Manager and the Structural Member Styles style type in the Style Manager.

If you are operating with the AIA layering system as your current layer standard, when you create members or convert AutoCAD entities to structural members using the menu picks or toolbars, AutoCAD Architecture assigns the new members to layers: A-Cols, A-Cols-Brce or A-Beam, respectively. If Generic AutoCAD Architecture is your standard, the layers used are A_Columns, A_Beams and A_Braces. If your layer standard is Current Layer, new entities come in on the current layer, as in plain vanilla AutoCAD.

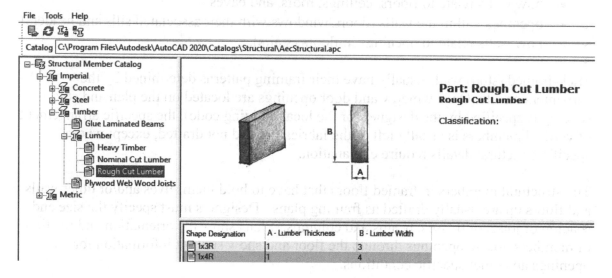

The Structural Member Catalog includes specifications for standard structural shapes. You can choose shapes from the Structural Member Catalog and generate styles for structural members that you create in your drawings.

> A style that contains the catalog shape that you selected is created. You can view the style in the Style Manager, create a new structural member from the style, or apply the style to an existing member.

> When you add a structural member to your drawing, the shape inside the style that you created defines the shape of the member. You define the length, justification, roll or rise, and start and end offsets of the structural member when you draw it.

You cannot use the following special characters in your style names:

- less-than and greater-than symbols (< >)
- forward slashes and backslashes (/ \)
- quotation marks (")
- colons (:)
- semicolons (;)
- question marks (?)
- commas (,)
- asterisks (*)
- vertical bars (|)
- equal signs (=)
- backquotes (`)

The left pane of the Structural Member Catalog contains a hierarchical tree view. Several industry standard catalogs are organized in the tree, first by imperial or metric units, and then by material.

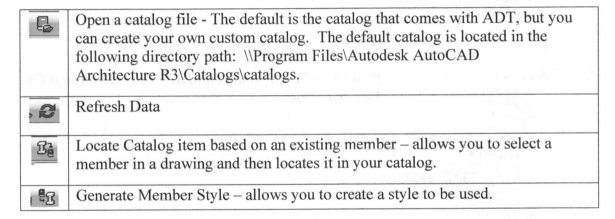

	Open a catalog file - The default is the catalog that comes with ADT, but you can create your own custom catalog. The default catalog is located in the following directory path: \\Program Files\Autodesk AutoCAD Architecture R3\Catalogs\catalogs.
	Refresh Data
	Locate Catalog item based on an existing member – allows you to select a member in a drawing and then locates it in your catalog.
	Generate Member Style – allows you to create a style to be used.

We will be adding a wood framed deck, 4000 mm x 2750 mm, to the back of the house. For purposes of this exercise we will assume that the ground level is 300 mm below the slab at the back of the house and falls away so that grade level below the edge of the deck away from the house is 2 meters below floor level: - 2 m a.f.f. (above finish floor) in architectural notation. We will place the top of the deck floorboards even with the top of the floor slab.

We will use support and rim joists as in standard wood floor framing, and they will all be at the same level, rather than joists crossing a support beam below. In practice this means the use of metal hangers, which will not be drawn. Once the floor system is drawn we will add support columns at the outside rim joist and braces at the columns.

Exercise 6-1:

Creating Member Styles

Drawing Name: member_styles.dwg
Estimated Time: 5 minutes

This exercise reinforces the following skills:

□ Creating Member Styles
□ Use of Structural Members tools

1.   Open *member_styles.dwg*.

2.   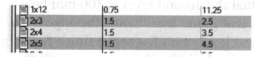 Activate the Manage ribbon.

Go to **Style & Display → Structural Member Catalog**.

3.   Browse to the **Imperial/Timber/Lumber/Nominal Cut Lumber** folder.

4.   In the lower right pane:

Locate the **2x4** shape designation.

1x12	0.75	11.25
2x3	1.5	2.5
2x4	1.5	3.5
2x5	1.5	4.5

5.   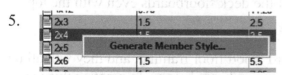 Right click and select **Generate Member Style**.

2x3	1.5	2.5
2x4		
2x5	Generate Member Style...	
2x6	1.5	5.5

6.   In the Structural Member Style dialog box, the name for your style automatically fills in.

Click **OK**.

7. 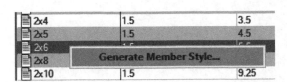 Locate the **2x6** shape designation. Right click and select **Generate Member Style**.

8. 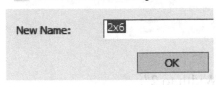 In the Structural Member Style dialog box, the name for your style is displayed.

Click **OK**.

9. Close the dialog.
Save as *ex6-1.dwg*.

Exercise 6-2:
Creating Member Shapes

Drawing Name: member_shapes.dwg
Estimated Time: 10 minutes

This exercise reinforces the following skills:

❑   Structural Member Wizard

1. 🗁   Open *member_shapes.dwg*.

2.  Activate the **Manage** ribbon.

Go to **Style & Display→Structural Member Wizard**.

3. Select the **Cut Lumber** category under Wood.

Click **Next**.

4. Set the Section Width to **2″**.

Set the Section Depth to **10″**.

Click **Next**.

5. Enter the style name as **2x10**.

Click **Finish**.

6. Activate the **Manage** ribbon.

Go to **Style & Display → Structural Member Wizard**.

7. Select the **Cut Lumber** category under Wood.

Click **Next**.

8. 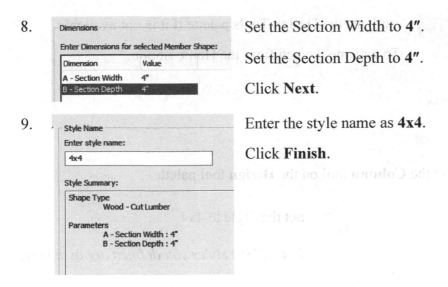 Set the Section Width to **4"**.

Set the Section Depth to **4"**.

Click **Next**.

9. Enter the style name as **4x4**.

Click **Finish**.

10. Save as *ex6-2.dwg*.

Exercise 6-3:
Adding Structural Members

Drawing Name: add_columns.dwg
Estimated Time: 25 minutes

This exercise reinforces the following skills:

- ❏ Creating Member Styles
- ❏ Use of Structural Members tools

1. 🔲 Open *add_columns.dwg*.

2. 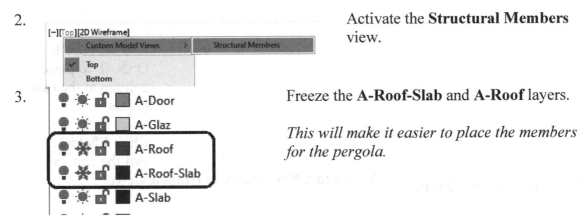 Activate the **Structural Members** view.

3. Freeze the **A-Roof-Slab** and **A-Roof** layers.

This will make it easier to place the members for the pergola.

4. 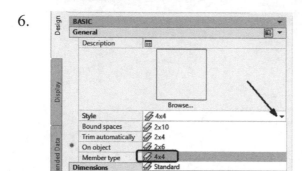 Activate the Design Tools palette if it is not available.

To activate, launch from the Home ribbon.

5. 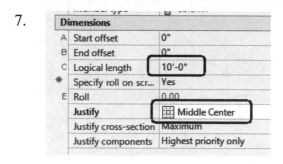 Select the **Column** tool on the **Design** tool palette.

6. 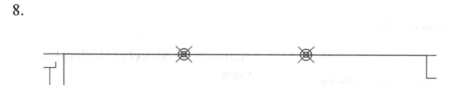 Set the Style to **4x4**.

Note all the styles you defined are available.

7. Set the Justify value to **Middle Center**.
Set the Logical Length to **10′ -0″**.

Dimensions	
A Start offset	0"
B End offset	0"
C Logical length	10'-0"
✳ Specify roll on scr...	Yes
E Roll	0.00
Justify	Middle Center
Justify cross-section	Maximum
Justify components	Highest priority only

8. Use the NODE OSNAP to place a column at the two points.

Click **ENTER** to exit the command.

9. [−][NW Isometric][Hidden]   Switch to a **NW Isometric Hidden** view.

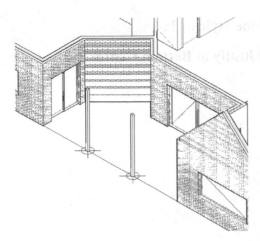

You should see two columns in the porch area.

10. **[-][Top][Hidden]**   Switch back to a **Top** view.

You can use the Zoom Previous tool to return to the previous view.

11.    Select the **Beam** tool from the Designs Palette.

12. 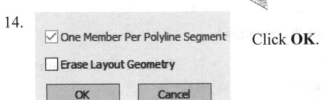   Right click and select **Apply Tool Properties to→Linework**.

13. 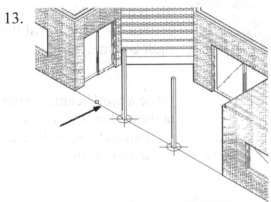   Select the red line.

Click **ENTER**.

14.   Click **OK**.

15.

Style	2x4
Bound spaces	By style (No)
Trim automatically	By style (Yes)
On object	Yes
Member type	Beam
Dimensions	
A Start offset	0"
B End offset	0"
C Logical length	1"
E Roll	0.00
Layout type	Fill
Justify	Bottom Center
Justify cross-section	Maximum
Justify components	Highest priority only

Set the Style to **2x4**.

Set Justify to **Bottom Center**.

16.

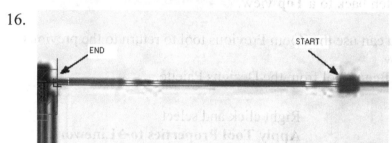

Select the center of the column as the start point and perpendicular point at the left wall as the end point.

17.

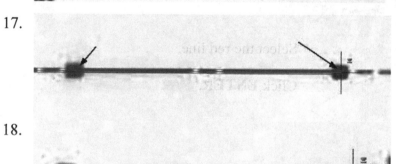

Place a second beam from the center of the left column to the center of the right column.

18.

Place a third beam starting at the node center of the column and perpendicular to the right wall.

19. Right click and select ENTER to exit the command.

20.

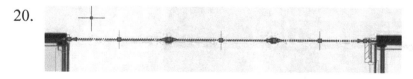

Select all three beams.

21.

Location	
Start point X	27'-6 7/32"
Start point Y	64'-0 1/16"
Start point Z	10'-0"
End point X	52'-1 31/32"
End point Y	64'-0 1/16"
End point Z	10'-0"
Rotation	0.00
Elevation	10'-0"
	🗐 Additional information

On the Properties palette:

Set the Start point Z to **10′ 0″**.
Set the End point Z to **10′ 0″**.

Note the Elevation updates to 10′ 0″.

Click ESC to release the selection set.

22.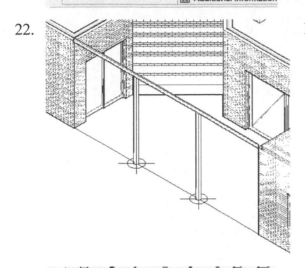

Inspect the columns and beams.

23.

Unsaved Layer State	▼
💡 ☀ 🔓 ⬛ 0	▼
💡 ☀ 🔓 ⬛ 0	
💡 ☀ 🔓 ⬛ A-Anno-Scrn	
💡 ☀ 🔓 ⬜ A-Door	
💡 ☀ 🔓 ⬜ A-Glaz	
💡 ☀ 🔓 ⬛ A-Roof	
💡 ☀ 🔓 ⬛ A-Roof-Slab	
💡 ☀ 🔓 ⬛ A-Slab	
💡 ☀ 🔓 ⬛ A-Wall	
💡 ☀ 🔓 ⬜ A-Wall-Open	
💡 ❄ 🔓 ⬛ column placement	
💡 ☀ 🔓 ⬛ Defpoints	

Set 0 as the current layer.

Thaw the A-Roof and A-Roof-Slab Layers.

Freeze the column placement layer.
This is the layer the points and line were on.

24. Save as *ex6-3.dwg*.

Notes:

QUIZ 6

True or False

1. Columns, braces, and beams are created using Structural Members.

2. When you isolate a Layer User Group, you are freezing all the layers in that group.

3. You can isolate a Layer User Group in ALL Viewports, a Single Viewport, or a Selection Set of Viewports.

4. When placing beams, you can switch the justification in the middle of the command.

5. Standard AutoCAD commands, like COPY, MOVE, and ARRAY, cannot be used in ACA.

Multiple Choice

6. Before you can place a beam or column, you must:

    A. Generate a Member Style
    B. Activate the Structural Member Catalog
    C. Select a structural member shape
    D. All of the above

7. Select the character that is OK to use when creating a Structural Member Style Name.

    A. -
    B. ?
    C. =
    D. /

8. Identify the tool shown.

    A. Add Brace
    B. Add Beam
    C. Add Column
    D. Structural Member Catalog

9. Setting a Layer Key

    A. Controls which layer AEC objects will be placed on.
    B. Determines the layer names created.
    C. Sets layer properties.
    D. All of the above.

ANSWERS:

1) T; 2) F; 3) T; 4) T; 5) F; 6) D; 7) A; 8) B; 9) A

Lesson 7:
Floors, Ceilings and Spaces

Slabs are used to model floors and other flat surfaces. You start by defining a slab style which includes the components and materials which are used to construct the floor system.

You can create a floor slab by drawing a polyline and applying the style to the polyline or by selecting the slab tool and then drawing the boundary to be used by the slab.

You can define multi-story buildings using the Project Navigator. In the Navigator, you can specify the number of levels for the building.

Ceiling grids can be used to define drop ceilings.

You can also use spaces to define floors and ceilings. You can use the Space/Zone Manager to apply different styles to floors and ceilings.

There are two reasons why you want to include floors and ceilings in your building model. 1) for your construction documentation and 2) for any renderings.

Exercise 7-1:
Creating Slab Styles

Drawing Name: slab_styles.dwg
Estimated Time: 30 minutes

This exercise reinforces the following skills:

- ❏ Creating Styles
- ❏ Copying Styles
- ❏ Renaming Styles
- ❏ Use of Materials

1.  Activate the Manage ribbon.

   Select **Style Manager**.

2. 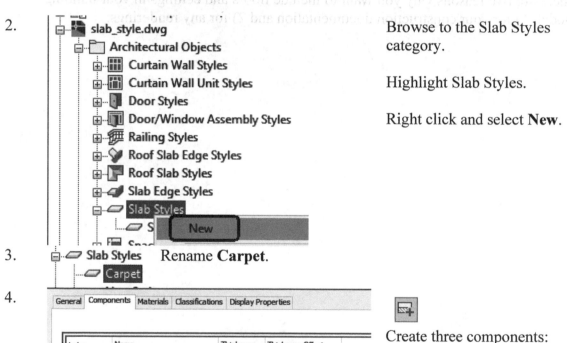 Browse to the Slab Styles category.

   Highlight Slab Styles.

   Right click and select **New**.

3. Rename **Carpet**.

4. Create three components:

   - Carpet
   - Plywood
   - Concrete Slab

   Set the carpet at 1/8" and ¼" Thickness Offset.
   Set the plywood at ¼" and 0" Thickness Offset.
   Set the concrete slab at 2" and -2" Thickness Offset.

5. Use the preview window to verify that the thicknesses and offsets are correct.

6.

Component	Material Definition
carpet	carpet
plywood	plywood sheathing
concrete slab	concrete - gray- in-place
Shrink Wrap Body	Standard

Create new materials for each of the components.

I have pre-loaded several materials in the exercise files. Some of the materials use image files that are part of the exercise files.

7. Edit the carpet material definition to use the **Berber – Pattern 2** render material.

8. Edit the plywood material definition to use the **Wood Plywood** render material.

9.

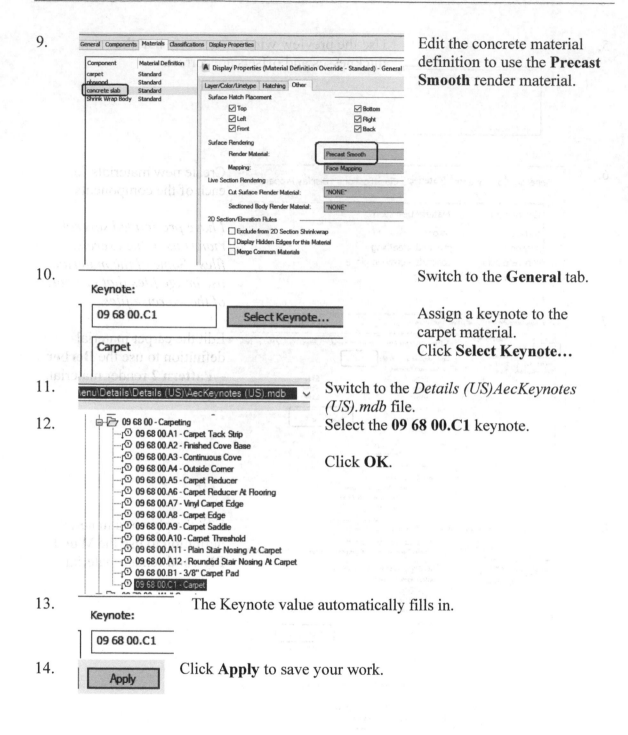

Edit the concrete material definition to use the **Precast Smooth** render material.

10. Switch to the **General** tab.

Assign a keynote to the carpet material.
Click Select Keynote…

11. Switch to the *Details (US)AecKeynotes (US).mdb* file.

12. Select the **09 68 00.C1** keynote.

Click **OK**.

13. The Keynote value automatically fills in.

14. Click **Apply** to save your work.

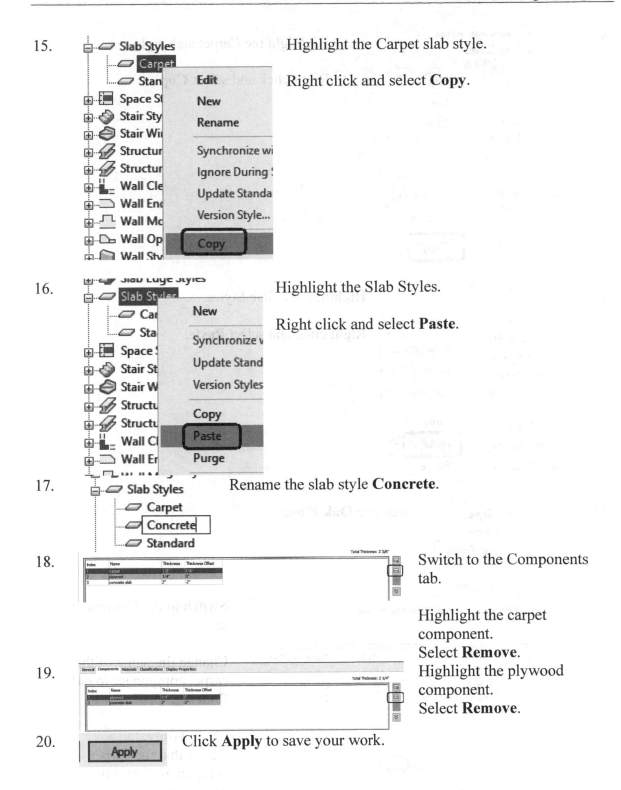

15. Highlight the Carpet slab style.

Right click and select **Copy**.

16. Highlight the Slab Styles.

Right click and select **Paste**.

17. Rename the slab style **Concrete**.

18. Switch to the Components tab.

Highlight the carpet component.
Select **Remove**.

19. Highlight the plywood component.
Select **Remove**.

20. Click **Apply** to save your work.

21.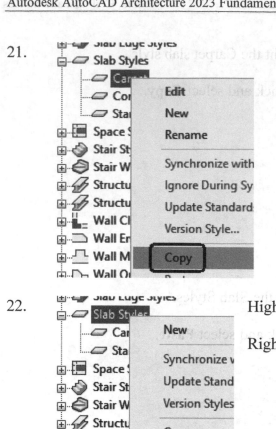

Highlight the Carpet slab style.

Right click and select **Copy**.

22.

Highlight the Slab Styles.

Right click and select **Paste**.

23.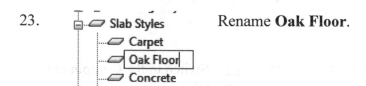

Rename **Oak Floor**.

24.

| General | Components | Materials | Classifications | Display Properties |

Index	Name	Thickness	Thickness Offset
1	oak flooring	1/8"	1/4"
2	plywood	1/4"	0"
3	concrete slab	2"	-2"

Switch to the Components tab.

Change the name of the carpet component to **oak flooring.**

25.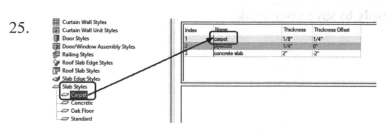

Click **Apply.**

Verify that the carpet component is still the top component for the carpet slab style.

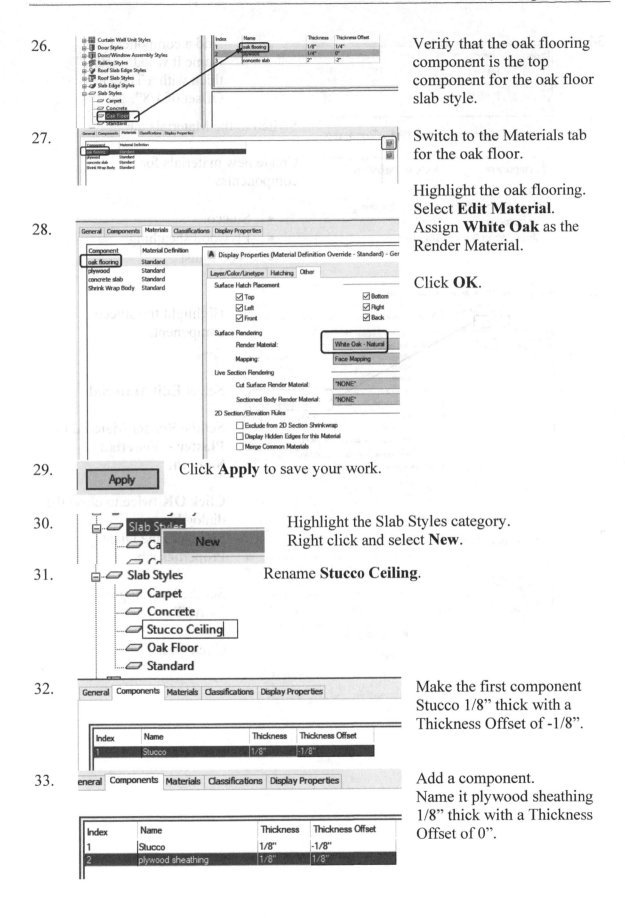

26. Verify that the oak flooring component is the top component for the oak floor slab style.

27. Switch to the Materials tab for the oak floor.

Highlight the oak flooring. Select **Edit Material**. Assign **White Oak** as the Render Material.

28. Click **OK**.

29. Click **Apply** to save your work.

30. Highlight the Slab Styles category. Right click and select **New**.

31. Rename **Stucco Ceiling**.

32. Make the first component Stucco 1/8" thick with a Thickness Offset of -1/8".

33. Add a component. Name it plywood sheathing 1/8" thick with a Thickness Offset of 0".

34.

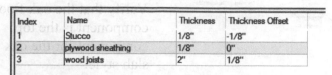

Index	Name	Thickness	Thickness Offset
1	Stucco	1/8"	-1/8"
2	plywood sheathing	1/8"	0"
3	wood joists	2"	1/8"

Add a component. Name it wood joists 2" thick with a Thickness Offset of 1/8".

35.

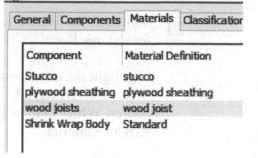

Switch to the Materials tab.

Create new materials for the components:

- Stucco
- Plywood sheathing
- Wood joist

36.

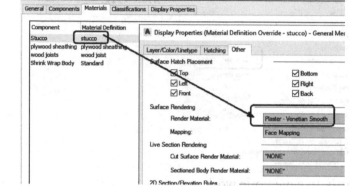

Highlight the stucco component.

Select **Edit Material**.

Set the Render Material to **Plaster – Venetian Smooth**.

Click **OK** twice to close the dialog boxes.
Switch to the Display Properties tab.

37.

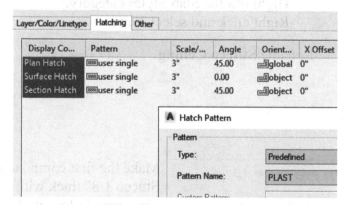

Set the Hatching to **PLAST**.

Click **OK**.

38.

General	Components	Materials	Classifications

Component	Material Definition
Stucco	stucco
plywood sheathing	plywood sheathing
wood joists	wood joist
Shrink Wrap Body	Standard

Highlight the plywood sheathing component.

Select **Edit Material**.

39.

Surface Rendering

Render Material: | Wood Plywood

Mapping: | Face Mapping

Select **Edit Display Properties**.
Select the Other tab.

Set the Render Material to **Wood Plywood**.
Click **OK** twice.

40.

Apply

Click **Apply** to save your work.

41.

Slab Edge Styles
Slab Styles
    Carpet
    Concrete
    Oak Floor
    Standard
    Stucco Ceiling

Space Styles
Stair Styles
Stair Winder
Structural Me
Structural Me
Wall Cleanup
Wall Endcap
Wall Modifier
Wall Opening
Wall Styles
Window Style

Edit
New
Rename
Synchronize with
Ignore During Sy
Update Standard
Version Style...
Copy

Highlight the **Stucco Ceiling** slab style.

Right click and select **Copy**.

42.

Slab Edge Styles
Slab Styles
    Ca
    Co
    Oa
    Sta
    Stu
    Stu
Space
Stair St
Stair W

New
Synchronize with Project Standards...
Update Standards from Drawing...
Version Styles...
Copy
Paste
Purge

Highlight the Slab Styles.

Right click and select **Paste**.

43. Rename **Acoustical Tile Ceiling**.

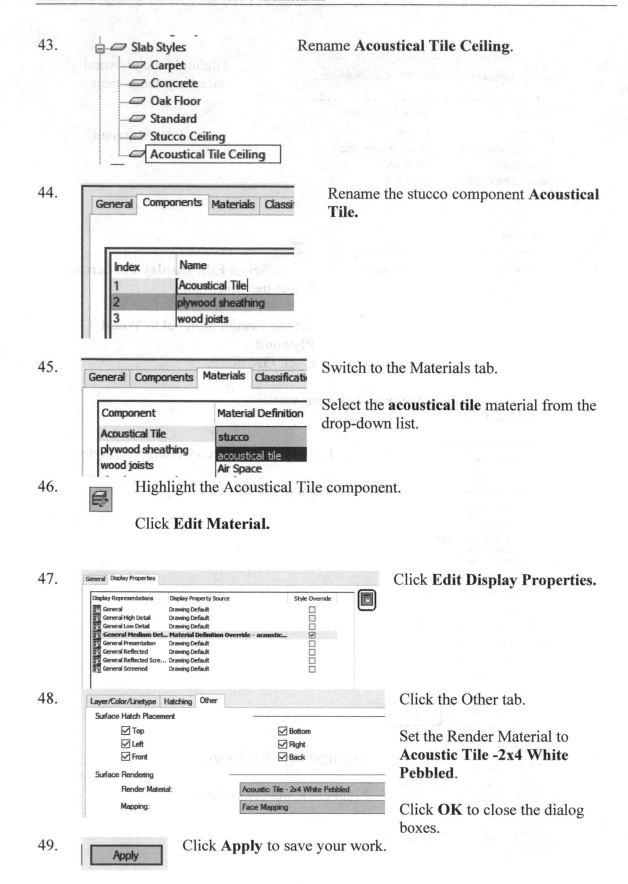

44. Rename the stucco component **Acoustical Tile.**

45. Switch to the Materials tab.

Select the **acoustical tile** material from the drop-down list.

46. Highlight the Acoustical Tile component.

Click **Edit Material.**

47. Click **Edit Display Properties.**

48. Click the Other tab.

Set the Render Material to **Acoustic Tile -2x4 White Pebbled**.

Click **OK** to close the dialog boxes.

49. Click **Apply** to save your work.

50.

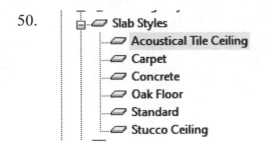

You should have the following slab styles defined:

- Acoustical Tile Ceiling
- Carpet
- Concrete
- Oak Floor
- Stucco Ceiling

Click **OK** to close the dialog.

Exercise 7-2:
Create Slab from Linework

Drawing Name: create_slab1.dwg
Estimated Time: 5 minutes

This exercise reinforces the following skills:

    ❑   Creating Slabs
    ❑   Named Views
    ❑   Polylines

1.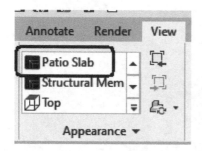

Activate the View ribbon.

Select the **Patio Slab** view to activate.

This will zoom into the patio area where the columns are located.

There is a red polyline to be used for the slab outline.

2.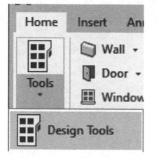

Click the Home ribbon.

Launch the **Design Tools**.

3.

Locate the Slab tool.

Right click and select **Apply Tool Properties to →Linework and Walls.**

4.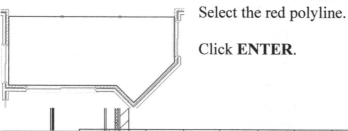

Select the red polyline.

Click **ENTER**.

5. Select **Yes**.

`]▼ SLABTOOLTOLINEWORK Erase layout geometry? [Yes No] <No>:`

6. Select **Direct**.

SLABTOOLTOLINEWORK Creation mode [Direct Projected] <Projected>:

7. Set the Justification to **Top**.

SLABTOOLTOLINEWORK Specify slab justification [Top Center Bottom Slopeline] <Bottom>:

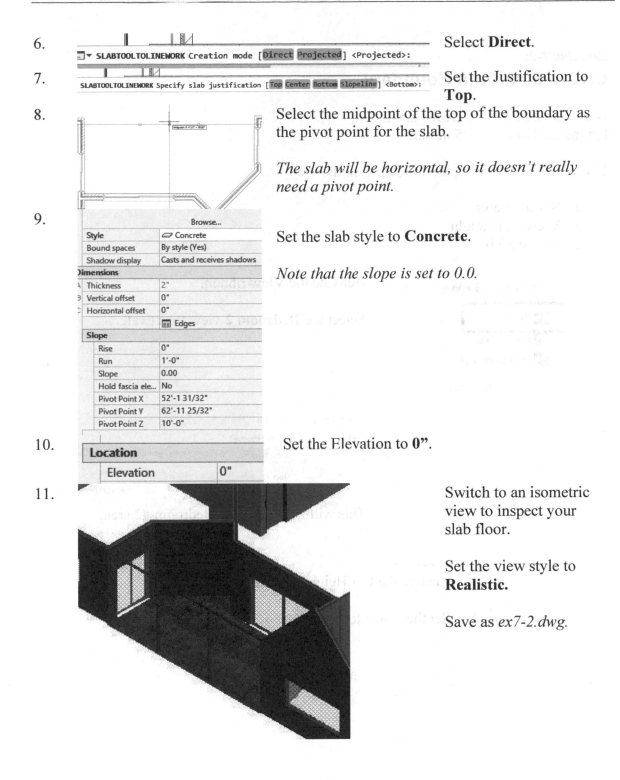

8. Select the midpoint of the top of the boundary as the pivot point for the slab.

The slab will be horizontal, so it doesn't really need a pivot point.

9. Set the slab style to **Concrete**.

Note that the slope is set to 0.0.

	Browse...	
Style	Concrete	
Bound spaces	By style (Yes)	
Shadow display	Casts and receives shadows	
Dimensions		
A Thickness	2"	
B Vertical offset	0"	
C Horizontal offset	0"	
	Edges	
Slope		
Rise	0"	
Run	1'-0"	
Slope	0.00	
Hold fascia ele...	No	
Pivot Point X	52'-1 31/32"	
Pivot Point Y	62'-11 25/32"	
Pivot Point Z	10'-0"	

10. Set the Elevation to **0"**.

Location	
Elevation	0"

11. Switch to an isometric view to inspect your slab floor.

Set the view style to **Realistic.**

Save as *ex7-2.dwg*.

Exercise 7-3:

Create Slab from Boundary

Drawing Name: create_slab2.dwg
Estimated Time: 5 minutes

This exercise reinforces the following skills:

- ❑ Saved Views
- ❑ View Cut Height
- ❑ Create Slab

1. 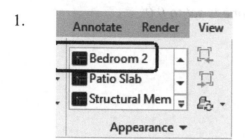   Activate the View ribbon.

   Select the **Bedroom 2** view to activate.

2.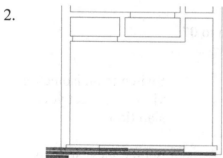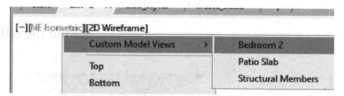

   This will zoom into the Bedroom #2 area.

3.   Change the Cut Height to adjust the view depth.

   Set the Value to **3'-0"**.

   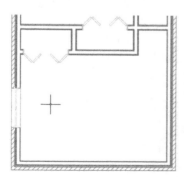

   The view updates.

4. Select the **Slab** tool from the Home ribbon.

We will add a carpet slab to this room.

5. Set the Style to **Carpet**.

6. *You are going to trace the boundary for the floor to be placed.*

Start at the upper left corner of the room.

7. Trace around the room.

At the last segment, right click and select **Close**.

Click ENTER when the boundary is completed.

8. If you hover over the boundary, you will see that you placed a slab using the Carpet style.

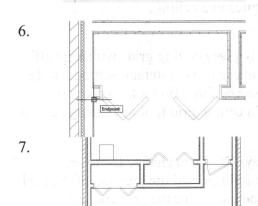

9.

Switch to an isometric view and inspect the floor you just placed in the bedroom area.

Save as *ex7-3.dwg*.

Ceiling Grids

Grids are AEC objects on which you can anchor other objects, such as columns, and constrain their locations. Grids are useful in the design and documentation phases of a project. A ceiling grid allows you to mount light fixtures to a ceiling.

The Design Tools palette contains tools which quickly place ceiling grids with a specific ceiling grid style and other predefined properties. You can use the default settings of the tool, or you can change its properties. You can also use ceiling grid tools to convert linework to ceiling grids and to apply the settings of a ceiling grid tool to existing ceiling grids.

You can change existing ceiling grids in different ways. You can change the overall dimensions of the grid, the number and position of grid lines, and the location of the grid within the drawing. You can also specify a clipping boundary for the grid and use that to mask the grid or insert a hole into the grid.

Exercise 7-4:

Create Ceiling Grid

Drawing Name: ceiling_grid.dwg
Estimated Time: 15 minutes

This exercise reinforces the following skills:

- ❑ Named Views
- ❑ Ceiling Grid
- ❑ Properties
- ❑ Set Boundary

1.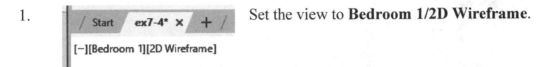
    Set the view to **Bedroom 1/2D Wireframe**.

2.
    Switch to the Home ribbon.
    Select **Ceiling Grid**.

3.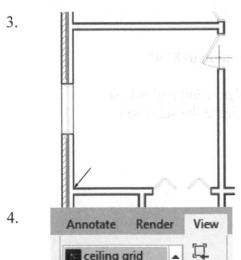
    Select the inside corner of the bedroom.

    Click **ENTER** to accept the default rotation of 0.00.

    Click **ESC** to exit the command.

4.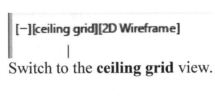
    Switch to the **ceiling grid** view.

5.   *The ceiling grid was placed at the 0 elevation.*

      Select the ceiling grid.

6.   Right click and select **Properties.**

Repeat '-VIEW
Recent Input
Isolate Objects
Basic Modify Tools
Clipboard
Paste From Content Browser
AEC Modify Tools
X Axis
Y Axis
Clip
Resize
Add Selected
AEC Dimension
Select Component
Edit Object Display...
Object Viewer...
Select Similar
Count
Deselect All
Properties

36'-2 31/32", 4.79675

7.   Change the Elevation to **8' 0"**.

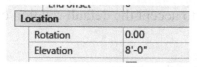

Location	
Rotation	0.00
Elevation	8'-0"

The ceiling grid position will adjust.
Click **ESC** to release the selection.

8. **[−][Bedroom 1][2D Wireframe]**

Switch back to the **Bedroom 1** view.

You won't see the ceiling grid because it is above the cut plane.

9. Select the **Rectangle** tool from the Draw panel on the Home ribbon.

10. Place a rectangle in the bedroom.

11. Select the rectangle.

End segment width	1/4"
Global width	1/4"
Elevation	0"

Set the Global Width to ¼"

This just makes it a little easier to see.

Click **ESC** to release the selection.

12.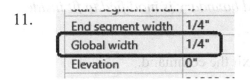

[−][ceiling grid][2D Wireframe]

Switch to the ceiling grid view.

13. Enable **Lineweight**.
Enable **Selection Cycling**.

14. 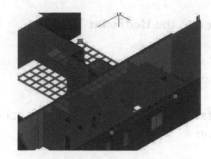 Select the ceiling grid.

15. 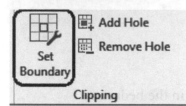 Select **Set Boundary** from the Clipping panel.

    Note: You won't see this tool unless the ceiling grid is selected.

16. 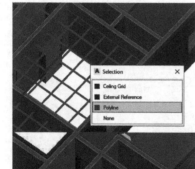 Click near the bottom edge of a wall.

    The Selection Cycling dialog should list the polyline that is the rectangle.

    Click on the Polyline in the list to select.

17.  *The ceiling grid boundary is modified to fit inside the room.*

    Click ESC to exit the command.

18. 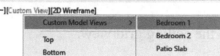 Switch back to a Bedroom 1 view.

19.

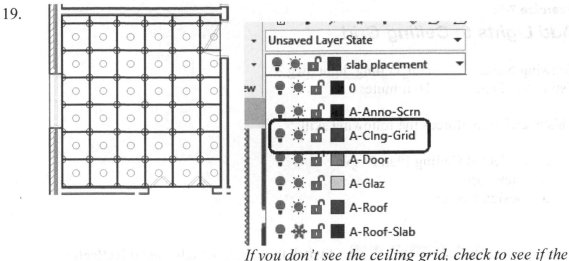

If you don't see the ceiling grid, check to see if the layer needs to be thawed.

Save as *ex7-4.dwg*.

Lighting

When there are no lights in a scene, the scene is rendered with default lighting. Default lighting is derived from one or two distant light sources that follow the viewpoint as you orbit around the model. All faces in the model are illuminated so that they are visually discernible. You can adjust the exposure of the rendered image, but you do not need to create or place lights yourself.

When you place user-defined lights or enable sunlight, you can optionally disable default lighting. Default lighting is set per viewport, and it is recommended to disable default lighting when user-defined lights are placed in a scene.

Light fixtures are placed for three reasons – 1) for rendering 2) to create lighting fixture schedules, so the construction documents can detail which light fixtures are placed where; and 3) for electrical wiring drawings.

Exercise 7-5:

Add Lights to Ceiling Grid

Drawing Name: add_ceiling_light.dwg
Estimated Time: 10 minutes

This exercise reinforces the following skills:

- ❑ Reflected Ceiling Plan Display
- ❑ Viewports
- ❑ Design Center

1.  Switch to the Layout tab named **Reflected Ceiling.**

2. 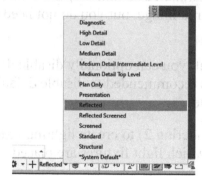 Click inside the viewport to activate Model space.

3.  On the status bar:

  Notice that the Display Configuration is set to **Reflected.**

4.  Switch to the View ribbon.

  Launch the **Design Center**.

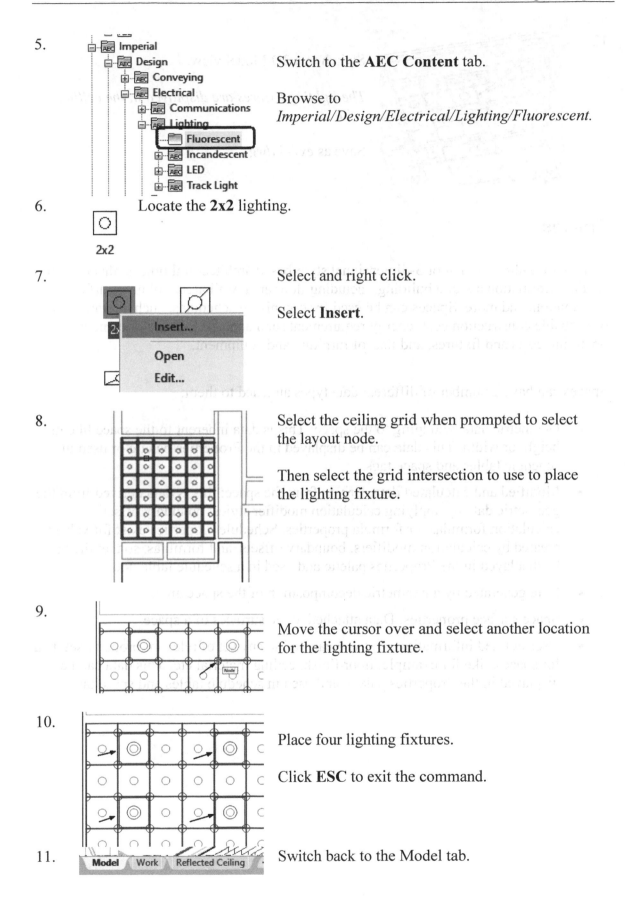

5. Switch to the **AEC Content** tab.

   Browse to
   Imperial/Design/Electrical/Lighting/Fluorescent.

6. Locate the **2x2** lighting.

7. Select and right click.

   Select **Insert**.

8. Select the ceiling grid when prompted to select the layout node.

   Then select the grid intersection to use to place the lighting fixture.

9. Move the cursor over and select another location for the lighting fixture.

10. Place four lighting fixtures.

    Click **ESC** to exit the command.

11. Switch back to the Model tab.

12.

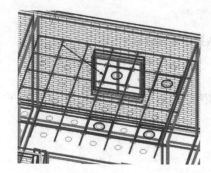

Switch to a 3D Model view.

The lighting fixtures are displayed in the ceiling grid.

Save as *ex7-5.dwg*.

Spaces

Spaces are 2-dimensional or 3-dimensional style-based architectural objects that contain spatial information about a building, including floor area, wall area, volume, surface information, and more. Spaces can be used for organizing schedules, such as statements of probable construction cost, energy requirements and analysis, leasing documents, operating costs and fixtures, and lists of furniture and equipment.

Spaces can have a number of different data types attached to them:

- Geometric data belonging to the space: This is data inherent to the space like its height or width. This data can be displayed in the Properties palette or used in schedule tables and space tags.

- Modified and calculated data derived from the space: this is data derived from the geometric data by applying calculation modifier styles, boundary offset calculation formulas, or formula properties. Schedule properties exist for values created by calculation modifiers, boundary offsets, and formulas, so that they can be displayed in the Properties palette and used in a schedule table, too.

- Data generated by a geometric decomposition of the space area.

- Space surface properties: Data attached to the surfaces of a space.

- User-defined information sets: you can define any set of relevant property set data for spaces, like for example, floor finish, ceiling material etc. This data can be displayed in the Properties palette and used in schedule tables and space tags.

Exercise 7-6:
Add a Space

Drawing Name: space.dwg
Estimated Time: 5 minutes

This exercise reinforces the following skills:

- ❑ Named Views
- ❑ Space

1. 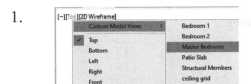 Switch to the **Master Bedroom** view.

2.  Select the **Space** tool from the Build panel on the Home ribbon.

3. 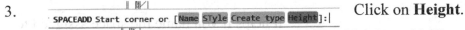 Click on **Height**.

4. ▼ SPACEADD Ceiling Height <8'-0">: 8'-2"| Set the Ceiling Height to **8'-2"**.

5.   Draw a rectangle in the top bedroom by selecting the opposing corners.

    Click **ESC** to exit the command.

6.

Switch to an isometric view so you can see the space.

You may need to click on the space to be able to see it.

7.

Rotate the view so you can see the underside of the space and the model.

Save as *ex7-6.dwg*.

Ceilings and Floors

You can use the Space/Zone Manager to modify the style that is applied to the floor or ceiling of a space. This will automatically update in any existing schedules.

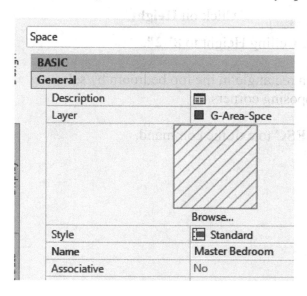

It helps if you name the space assigned to each room before you start applying different styles to ensure that you apply the correct styles to each space.

You can create space styles that use specific floor and ceiling materials to make it more efficient when defining your spaces.

Exercise 7-7:
Add a Ceiling and Floor to a Space

Drawing Name: ceiling_space.dwg
Estimated Time: 10 minutes

This exercise reinforces the following skills:

- ❑ Add Slab
- ❑ Slab Styles
- ❑ Properties

1.   -][SW Isometric][Hidden]   Switch to a SW Isometric view.

2.   Launch the **Design Tools**.

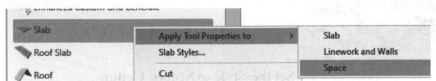

3.   Locate the Slab tool.
Right click and select **Apply Tool Properties to→ Space.**

4.

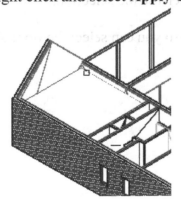

Select the space that is located in the upper bedroom.

Click **ENTER** to complete selection.

5.

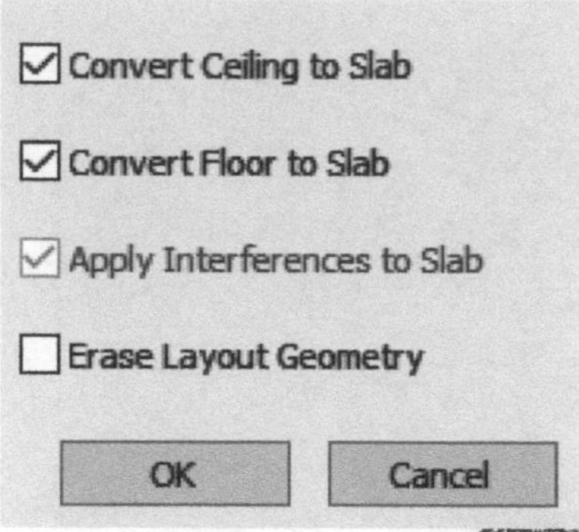
Convert Ceiling to Slab
Convert Floor to Slab
Apply Interferences to Slab
Erase Layout Geometry
OK
Cancel

Enable **Convert Ceiling to Slab**.
Enable **Convert Floor to Slab.**

Click **OK**.

6.

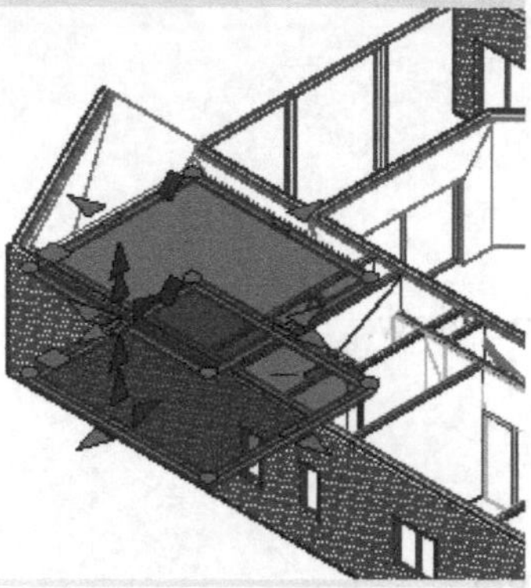

The two slabs appear.

Click **ESC** to release the selection.

7.

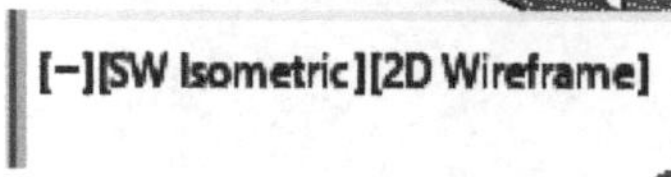
[–][SW Isometric][2D Wireframe]

Switch to a **2D Wireframe** display.

8.

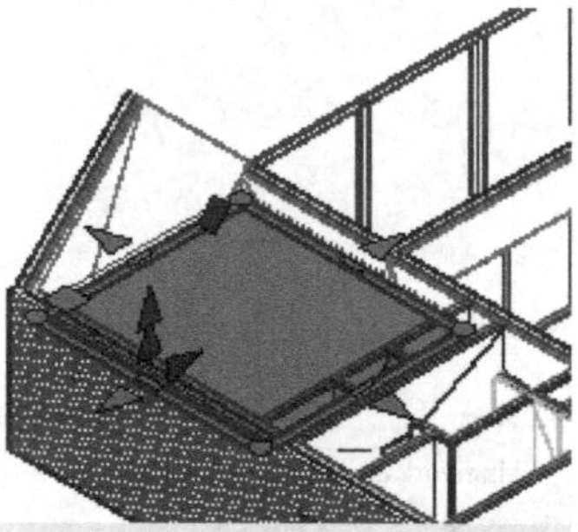

Select the ceiling slab.

9.

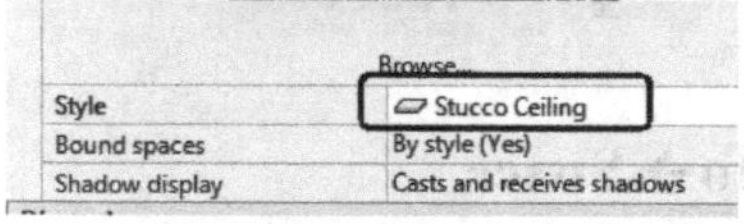
Browse...
Style
Stucco Ceiling
Bound spaces
By style (Yes)
Shadow display
Casts and receives shadows

Set the Style to **Stucco Ceiling**.

Click **ESC** to release the selection.

10.

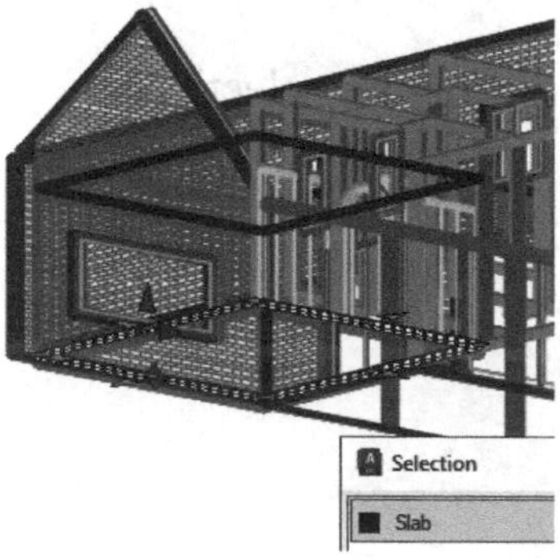
Selection
Slab

Orbit the view so you can select the floor slab.

11. 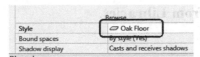 Set the Style to **Oak Floor**.

Click **ESC** to release the selection.

12. Save as *ex7-7.dwg*.

Exercise 7-8:
Add a Lighting Fixture to a Ceiling

Drawing Name: lighting fixture.dwg
Estimated Time: 5 minutes

This exercise reinforces the following skills:

- ❑ Inserting Blocks
- ❑ Properties

1. [−][Top][Realistic]   Set the view to **Top Realistic**.

2. 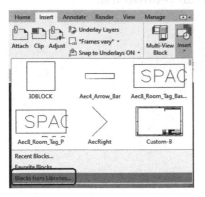 Switch to the Insert ribbon.

Select **Insert →Blocks from Libraries**.

3.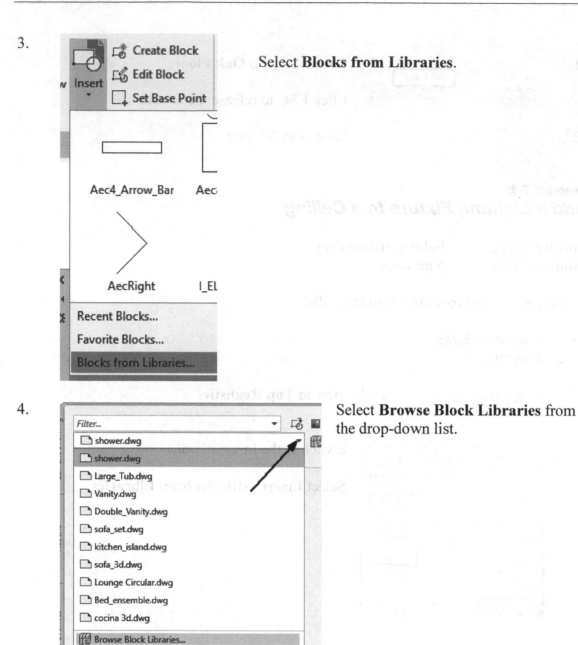

Select **Blocks from Libraries**.

4. Select **Browse Block Libraries** from the drop-down list.

5. Locate the *ceiling_fan.dwg* file located in the downloaded exercises.

    Click **Open**.

6. Highlight the file.

    Right click and select **Insert**.

7.

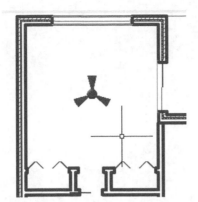

Place in the center of the master bedroom.

8.

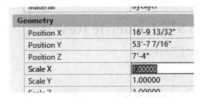

Select the fan and set the Position Z to **7'-4"**.

It took me some trial and error to figure out the correct Z value to place the fan.

9.

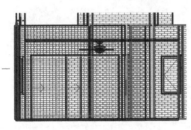

Switch to an isometric 2D Wireframe view and a Left view to inspect the placement of the ceiling fan.

Save as *ex7-8.dwg*.

Place in the center of the smaller bedroom.

Exercise 7-9:
Creating a Multi-Level Building

Drawing Name: multi-level.dwg
Estimated Time: 15 minutes

This exercise reinforces the following skills:

- ❑ Array
- ❑ Slabs

1.

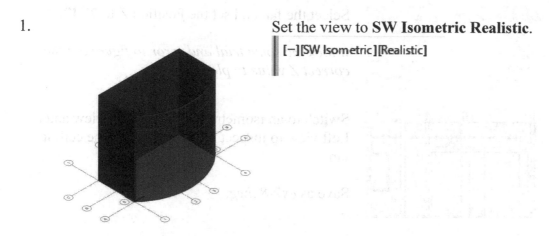

Set the view to **SW Isometric Realistic**.

[−][SW Isometric][Realistic]

2.

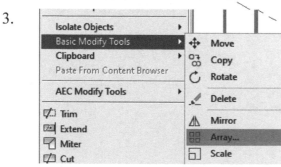

There is already a concrete slab placed on the first level.

Hover your mouse to locate the slab and select.

3.

Right click and select **Basic Modify Tools→Array.**

4.

ARRAY Enter array type [Rectangular PAth POlar] <Rectangular>:

Select **Rectangular**.

5.

Set the Columns to **1**.
Set the Rows to **1**.
Set the Levels to **3**.
Set the spacing between Levels to **3000**.

Disable **Associative.**

AEC elements cannot be used in an associative array.

Check the Preview to see the slabs that are placed.

Close
Array

Green check to accept the preview.

6.
💡 ☀ 🔓 ■ Linework

Thaw the Linework layer.

7.

If you zoom into the front corner of the building you should be able to locate a polyline that was used to place the floor slab.

We will use this polyline to create a ceiling slab.

Click **ESC** to release any selection.

Selection

Slab

Wall

Column Grid

Polyline

8.

Launch the Design Tools.

Locate the Slab tool on the Design tab.

Select **Slab→Apply Tool Properties to →Linework and Walls.**

9.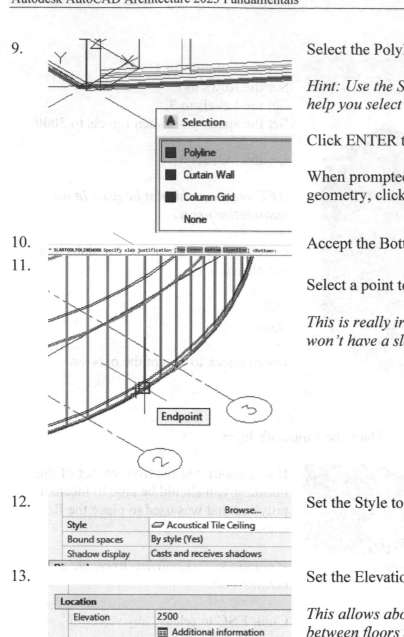

Select the Polyline.

Hint: Use the Selection Cycling tool to help you select the correct element.

Click ENTER to complete the selection.

When prompted to erase layout geometry, click No.

10. Accept the Bottom default.

11. Select a point to use as the pivot edge.

This is really irrelevant since our ceiling won't have a slope.

12. Set the Style to **Acoustical Tile Ceiling.**

	Browse...
Style	⬭ Acoustical Tile Ceiling
Bound spaces	By style (Yes)
Shadow display	Casts and receives shadows

13. Set the Elevation to **2500**.

Location	
Elevation	2500
	🖿 Additional information

This allows about 400 mm space between floors and ceilings for HVAC, electrical and plumbing elements.

14.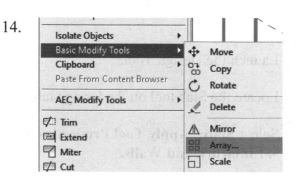

Right click and select **Basic Modify Tools→Array.**

15. ARRAY Enter array type [Rectangular PAth POlar] <Rectangular>:

Select **Rectangular**.

16.

Set the Columns to **1**.
Set the Rows to **1**.
Set the Levels to **4**.
Set the spacing between Levels to **3000**.

Disable **Associative**.

AEC elements cannot be used in an associative array.

Check the Preview to see the slabs that are placed.

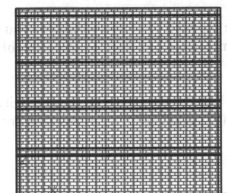

Close
Array

Green check to accept the preview.

17.

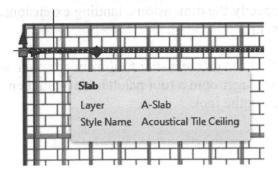

Switch to a Front view/2D Wireframe.

Slab	
Layer	A-Slab
Style Name	Acoustical Tile Ceiling

If you hover over a slab, you can see the style of slab applied.

18.

Save as *ex7-9.dwg*.

Stairs

Stairs are AEC objects that use flights of treads and risers to accommodate vertical circulation. Stairs also interact with railing objects. You can control the style of the stair, the shape of the landing, the type of treads, and the height and width of the stair run. By default, ACA will comply with standard business codes regarding the width of the tread and riser height. An error will display if you attempt to enter values that violate those standards.

You can modify the edges of stairs so that they are not parallel, and ACA allows curved stairs.

Landings can also be non-rectangular. Stairs allow the use of nearly arbitrary profiles for the edges of flights and landings. In addition, railings and stringers can be anchored to stairs and can follow the edges of flights and landings. You can create custom stairs from linework or profiles to model different conditions as well.

Materials can be assigned to a stair. These materials are displayed in the Realistic visual style, or when rendered. Materials have specific settings for the physical components of a stair, such as risers, nosing, and treads. You can include the material properties in any schedules.

Stair styles allow you to predefine materials for the stair. A stair style allows you to specify the dimensions, landing extensions, components, and display properties of the stair.

You can create a stair tool from any stair style. You can drag the style from the Style Manager onto a tool palette. You can then specify default settings for any stair created from the tool.

Exercise 7-10:

Placing a Stair

Drawing Name: stair.dwg
Estimated Time: 10 minutes

This exercise reinforces the following skills:

❑   Stair

1.

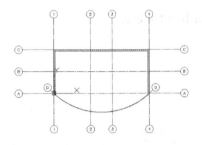

Set the view to **Top 2D Wireframe**.

[–][Top][2D Wireframe]

2.

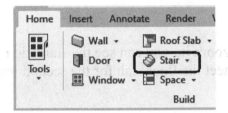

Select the **Stair** tool on the Home ribbon.

3.

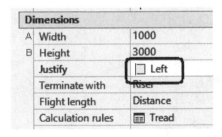

Set Justify to **Left**.

4.

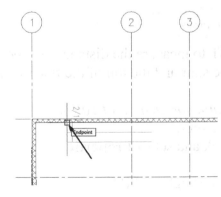

Select the endpoint of the construction line located between Grids 1 and 2 as the start point.

Draw to the right.

5. When all the risers are shown, click to end the stair run.

Click ENTER to exit the command.

6. Switch to a 3D view to inspect your stairs.

Switch to a front view.

7. If you zoom in, you can see the stairs do not quite meet the floor slab for the second level.

8.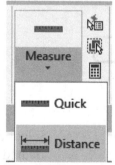

Use DIST to measure the distance between the top of the stair and the top of the floor slab.

9.

	Dimensions	
A	Width	1000
B	Height	3100
	Justify	☐ Left
	Terminate with	Riser

I got a value for Delta Z of 100 mm.
Select the stair.
Right click and select **Properties**.

Adjust the Height to **3100**.

10.

Click ESC to release the stair selection.

Verify that the top of the stair meets the top of the floor slab.

11. Save as *ex7-10.dwg*.

Railings

Railings are objects that interact with stairs and other objects. You can add railings to existing stairs, or you can create freestanding railings.

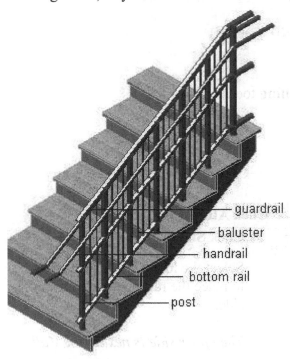

Railings can have guardrails, handrails, posts, balusters, and one or more bottom rails. Additionally, you can add custom blocks to railings.

A railing style can control the properties for all railings that use that style, rather than changing properties for each railing in the drawing. Various styles exist in the templates for common railing configurations, such as guardrail pipe, handrail grip, and handrail round. Within the railing style, you can specify the rail locations and height, post locations and intervals, components and profiles, extensions, material assignments, and display properties of the railing.

Exercise 7-11:
Placing a Railing

Drawing Name: stair_railing.dwg
Estimated Time: 5 minutes

This exercise reinforces the following skills:

❑   railing

1.   Set the view to **Top 2D Wireframe**.

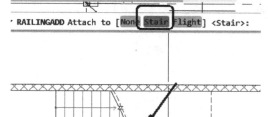

[−][Top][2D Wireframe]

2.   Select the **Railing** tool.

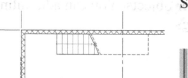

3.   Click **Attach**.

4.   Click **Stair**.

5.   Select the lower side of the stair to place the railing on the lower side.

The upper side is next to the wall.

Click **ENTER** to complete the stair command.

6.   Set the view to **SW Isometric Realistic**.

[−][SW Isometric][2D Wireframe]

Inspect the railing placed on the stair.

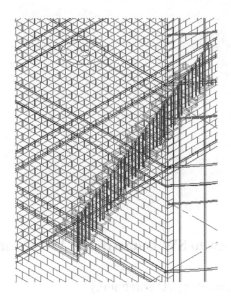

7. Save as *ex7-11.dwg*.

Exercise 7-12:
Placing a Stair Tower

Drawing Name: stair_tower.dwg
Estimated Time: 30 minutes

This exercise reinforces the following skills:

- Array
- Cut Hole in Slab

1.

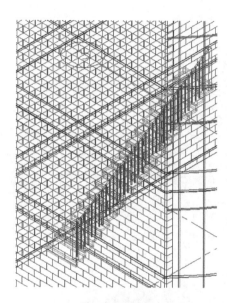

Set the view to **SW Isometric 2D Wireframe**.

[−][SW Isometric][2D Wireframe]

The Stair Tower tool requires that you place the stairs in a separate file and attach it to your floor plan as an external reference using a Construct. I have heard all sorts of issues using this method from the stairs not attaching properly to losing the reference, so this is not my best practice. Sometimes the simplest answer is the best one.

2.

Repeat Array...
Recent Input
Isolate Objects
Basic Modify Tools
Clipboard
Paste From Content Browser
AEC Modify Tools
AEC Dimension
Select Component
Edit Object Display...

Move
Copy
Rotate
Delete
Mirror
Array...
Scale

Select the **stairs** and **railing.**
Click ENTER to complete the selection.

Right click and select **Basic Modify Tools→Array.**

4.

ARRAY Enter array type [Rectangular Path POlar] <Rectangular>:

Select **Rectangular**.

5.

	Columns: 1		Rows: 1		Levels: 2			
Rectangular	Between: 19851		Between: 12854		Between: 3100		Associative Base Point	Close Array
	Total: 19851		Total: 12854		Total: 6253			
Type	Columns		Rows		Levels		Properties	Close

Set the Columns to **1**.
Set the Rows to **1**.
Set the Levels to **2**.
Set the spacing between Levels to **3100.**

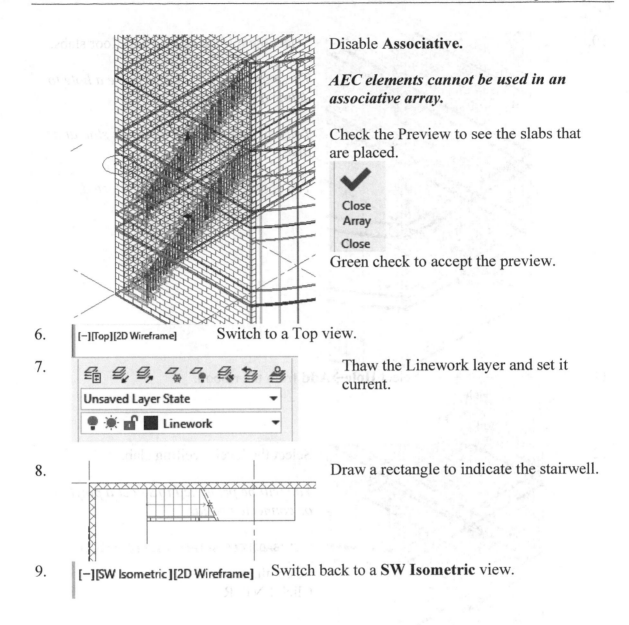

Disable **Associative.**

AEC elements cannot be used in an associative array.

Check the Preview to see the slabs that are placed.

Close
Array

Close

Green check to accept the preview.

6.　[−][Top][2D Wireframe]　Switch to a Top view.

7.　Unsaved Layer State

　　Linework

　　Thaw the Linework layer and set it current.

8.　Draw a rectangle to indicate the stairwell.

9.　[−][SW Isometric][2D Wireframe]　Switch back to a **SW Isometric** view.

10. Select the middle ceiling and floor slabs.

These are the slabs that require a hole to define a stairwell.

You can only add a hole to one slab at a time.

You should have four slabs selected.

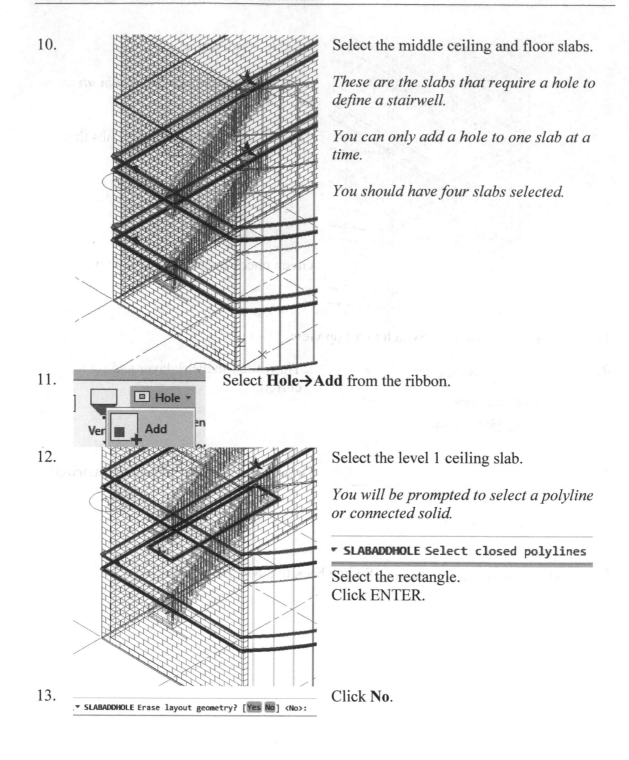

11. Select **Hole→Add** from the ribbon.

12. Select the level 1 ceiling slab.

You will be prompted to select a polyline or connected solid.

▾ SLABADDHOLE Select closed polylines

Select the rectangle.
Click ENTER.

13. ▾ SLABADDHOLE Erase layout geometry? [Yes No] <No>:

Click **No**.

14.

The hole is added to the level 1 ceiling.

Click **ESC** to release the selection.

15.

Select the level 2 floor slab.

Select **Hole→Add** from the ribbon.

16.

You will be prompted to select a polyline or connected solid.

▾ **SLABADDHOLE** Select closed polylines

Select the rectangle.
Click ENTER.

17.

▾ SLABADDHOLE Erase layout geometry? [Yes No] <No>:

Click **No**.

18.

You can see the added hole for the stairwell.

Click **ESC** to release the selection.

19. 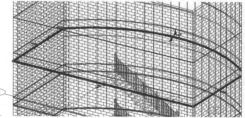 Select the level 2 ceiling slab.

Select **Hole→Add** from the ribbon.

20.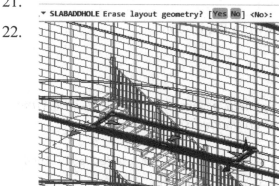

You will be prompted to select a polyline or connected solid.

▾ **SLABADDHOLE** Select closed polylines

Select the rectangle.
Click ENTER.

21. 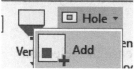 Click **No**.

▾ SLABADDHOLE Erase layout geometry? [Yes No] <No>:

22.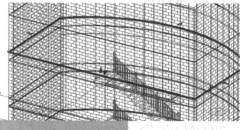

You can see the added hole for the stairwell.

Click **ESC** to release the selection.

23. 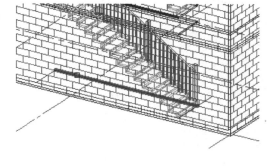 Select the Level 3 floor slab.

24. Select **Hole→Add** from the ribbon.

25.

You will be prompted to select a polyline or connected solid.

▾ **SLABADDHOLE** Select closed polylines

Select the rectangle.
Click ENTER.

26. Click **No**.

▾ SLABADDHOLE Erase layout geometry? [Yes No] <No>:

27.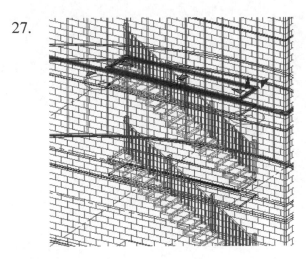

You can see the added hole for the stairwell.

Click **ESC** to release the selection.

28.   Save as *ex7-12.dwg*.

Notes:

QUIZ 7

True or False

1. You can create a floor or ceiling slab by converting a space.

2. Once you place a ceiling grid, it can't be modified.

3. When you place a lighting fixture on a ceiling grid, you cannot snap to the grid.

4. Railings are automatically placed on both sides of a selected stair.

5. You can assign materials to different components of a railing.

Multiple Choice

6. Slabs can be used to model: (select all that apply)

    A. Roofs
    B. Floors
    C. Ceilings
    D. Walls

7. Ceiling grids can be created using: (select all that apply)

    A. Linework
    B. Walls
    C. Roof Slabs
    D. Arcs

8. Space Styles: (select all that apply)

    A. Control the hatching displayed
    B. Control the name displayed when spaces are tagged
    C. Cannot be modified
    D. Have materials assigned

9. Railings can have the following: (select all that apply)

    A. Guardrails.
    B. Handrails.
    C. Posts.
    D. Balusters.

10. The first point selected when placing a set of stairs is:

    A. The foot/bottom of the stairs
    B. The head/top of the stairs
    C. The center point of the stairs
    D. Depends on the property settings

ANSWERS:

1) T; 2) F; 3) F; 4) F; 5) T; 6) D; 7) A, B, C; 8) A, B; 9) A, B, C, D; 10) A

Lesson 8:
Rendering & Materials

AutoCAD Architecture divides materials into TWO locations:

- Materials in Active Drawing
- Autodesk Library

You can also create one or more custom libraries for materials you want to use in more than one project. You can only edit materials in the drawing or in the custom library—Autodesk Materials are read-only.

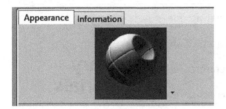

Materials have two asset definitions—Appearance and Information. Materials do not have to be fully defined to be used. Each asset contributes to the definition of a single material.

You can define Appearance independently without assigning/associating it to a material. You can assign the same assets to different materials. Assets can be deleted from a material in a drawing or in a user library, but you cannot remove or delete assets from materials in locked material libraries, such as the Autodesk Materials library or any locked user library. You can delete assets from a user library – BUT be careful because that asset might be used somewhere!

You can only duplicate, replace or delete an asset if it is in the drawing or in a user library. The Autodesk Asset Library is "read-only" and can only be used to check-out or create materials.

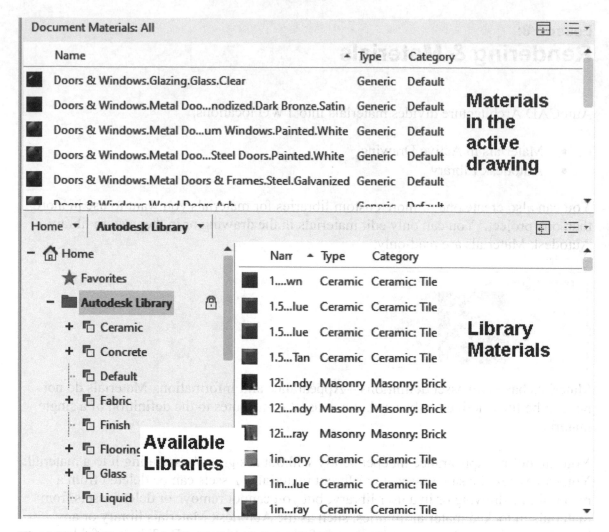

The Material Browser is divided into three distinct panels.

The Document Materials panel lists all the materials available in the drawing.

The Material Libraries list any libraries available.

The Library Materials lists the materials in the material libraries.

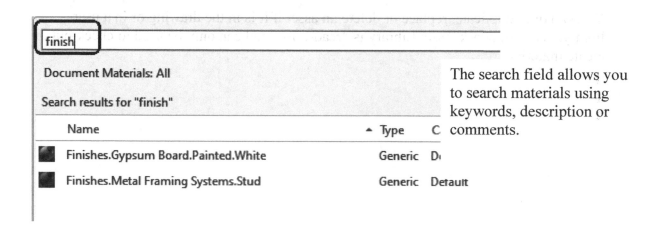

The search field allows you to search materials using keywords, description or comments.

The View control button allows you to control how materials are sorted and displayed in the Material Libraries panel.

Document Materials

✔ Show All

Show Applied

Show Selected

Show Unused

Purge All Unused

View Type

Thumbnail View

✔ List View

Text View

Sort

✔ by Name

by Type

by Material Color

by Category

Thumbnail Size

16 x 16

24 x 24

32 x 32

64 x 64

✔ 256 x 256

Material Editor

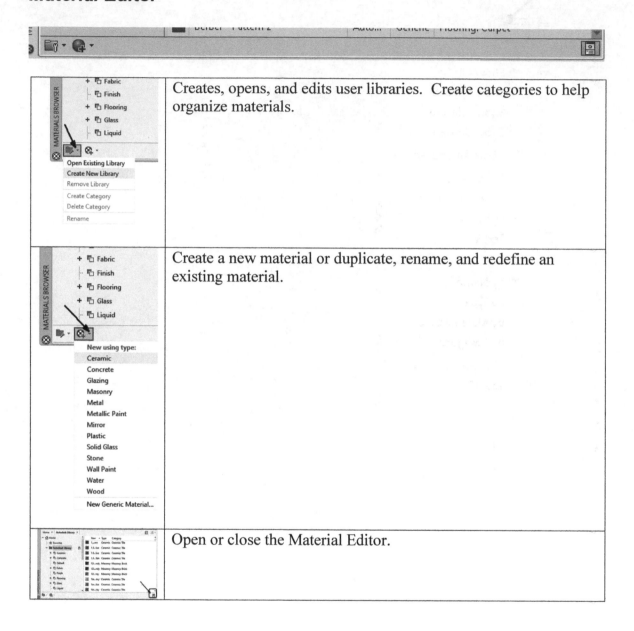

	Creates, opens, and edits user libraries. Create categories to help organize materials.
	Create a new material or duplicate, rename, and redefine an existing material.
	Open or close the Material Editor.

Exercise 8-1:
Modifying the Material Browser Interface

Drawing Name: material_browser.dwg
Estimated Time: 10 minutes

This exercise reinforces the following skills:

❑ Navigating the Material Browser and Material Editor

1.  Activate the **Render** ribbon.

   Select the small arrow in the lower right corner of the materials panel.

   This launches the Material Editor.

2. 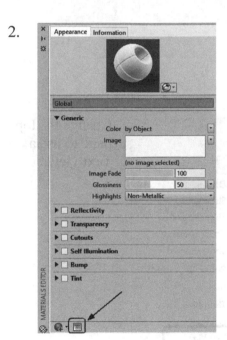 Launch the Material Browser by selecting the button on the bottom of the Material Editor dialog.

3.

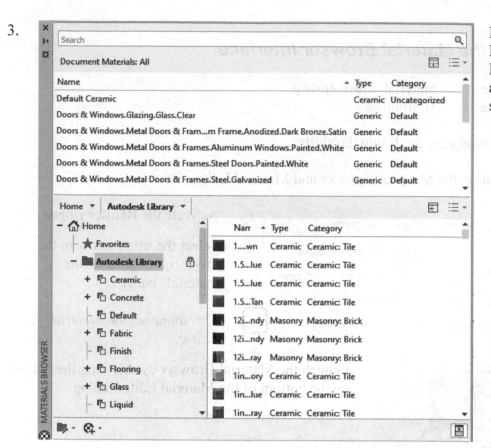

Make the Materials Browser appear as shown.

4.

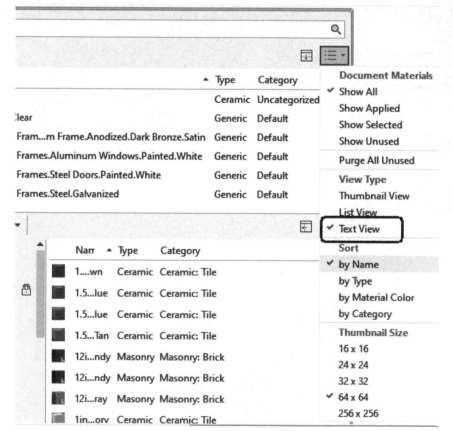

Change the top pane to use a Text view.

5.

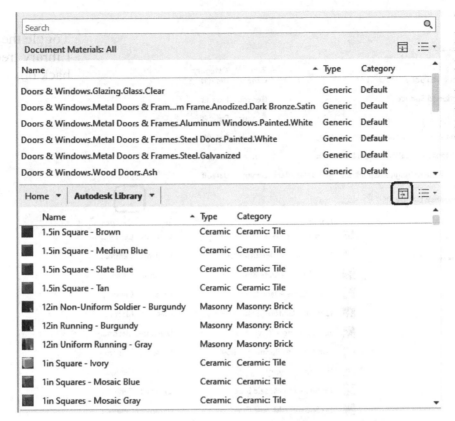

Toggle the Materials Library list OFF by selecting the top button on the Documents pane.

6.

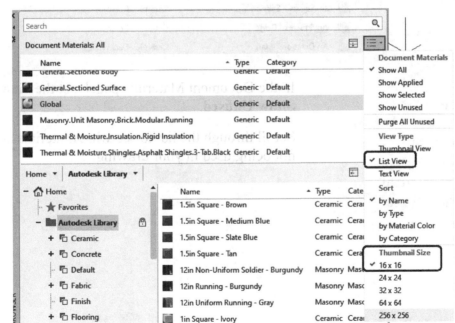

Change the upper pane to a **List View**.
Set the Thumbnail size to **16x16.**

7.

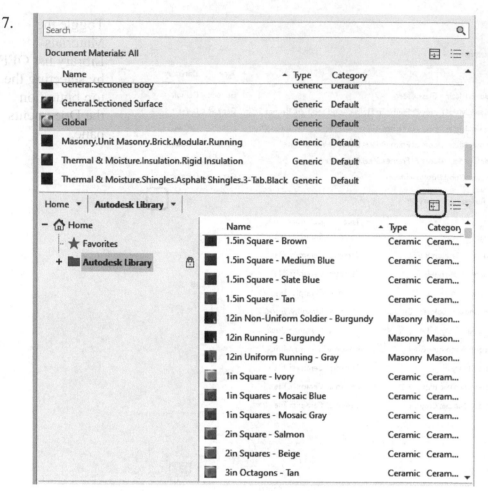

Toggle the Library tree back on.

8.

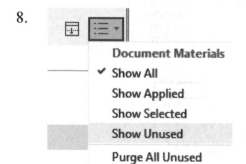

In the Document Materials panel, set the view to **Show Unused**.

Scroll through the list to see what materials are not being used in your drawing.

9.

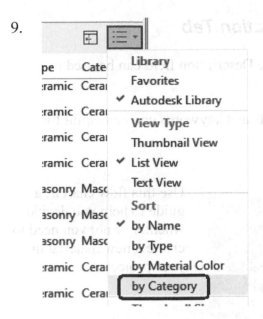

In the Material Library pane, change the view to sort **by Category**.

10. Close the Material dialogs.

11. Save as *ex8-1.dwg*.

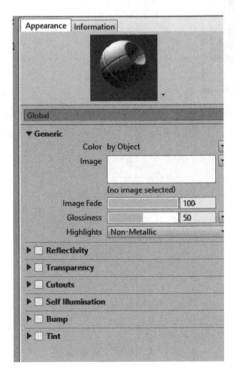

The Appearance Tab

Appearance controls how material is displayed in a rendered view, Realistic view, or Ray Trace view.

A zero next to the hand means that only the active/selected material uses the appearance definition.

Different material types can have different appearance parameters.

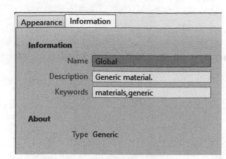

The Information Tab

The value in the Description field can be used in Material tags.

Using comments and keywords can be helpful for searches.

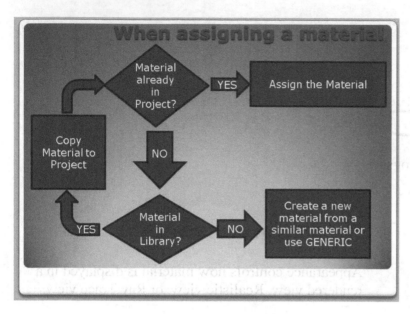

Use this flow chart as a guide to help you decide whether or not you need to create a new material in your project.

Exercise 8-2:
Copy a Material from a Library to a Drawing

Drawing Name: copy_materials.dwg
Estimated Time: 10 minutes

In order to use or assign a material, it must exist in the drawing. You can search for a material and, if one exists in the Autodesk Library, copy it to your active drawing.

This exercise reinforces the following skills:

- ❑ Copying materials from the Autodesk Materials Library
- ❑ Modifying materials in the drawing

1. Go to the **Render** ribbon.

2.    Select the **Materials Browser** tool.

3.   Type **yellow** in the search field.

There are no yellow materials in the drawing.

4.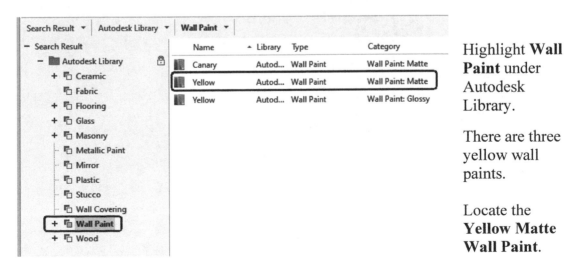

Highlight **Wall Paint** under Autodesk Library.

There are three yellow wall paints.

Locate the **Yellow Matte Wall Paint**.

5. 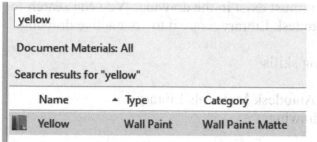 Select the up arrow to copy the material up to the drawing.

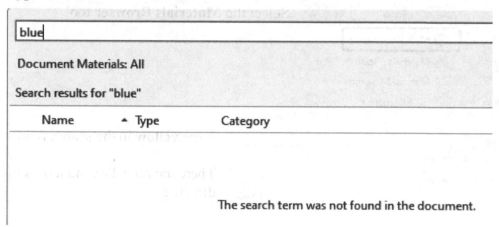

The yellow paint material is now listed in the Document Materials.

6. Type **blue** in the search field.

7. There are no blue materials in the drawing, but the Autodesk Library has several different blue materials available.

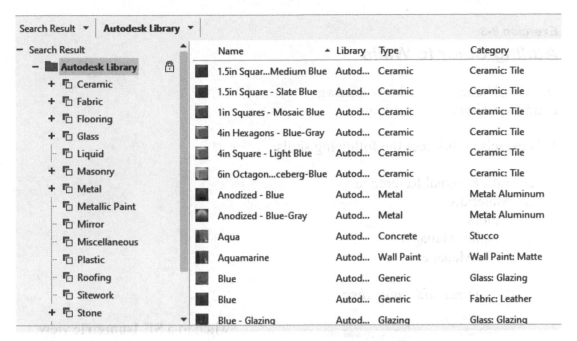

8.   Highlight the Wall Paint category in the Library pane.

This confines the search to blue wall paints.

9.   Locate the **Periwinkle Matte** Wall Paint and then copy it over to the drawing.

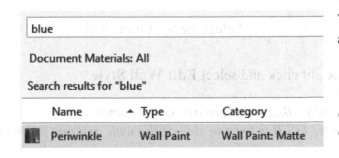

Notice the lock symbol next to the material name. Materials in the library cannot be modified. However, once you import a material into your document, it can be edited.

The blue wall paint is now available in the drawing.

You have added material definitions to the active drawing ONLY.

10.   Close the Materials dialog.

11.   Save as *ex8-2.dwg*.

Exercise 8-3:

Adding Color to Walls

Drawing Name: add_color.dwg
Estimated Time: 5 minutes

This exercise reinforces the following skills:

- ❑ Edit External Reference
- ❑ Materials
- ❑ Colors
- ❑ Display Manager
- ❑ Style Manager

1. Open *add_color.dwg*.

2.    Switch to a **SE Isometric** view.

3. 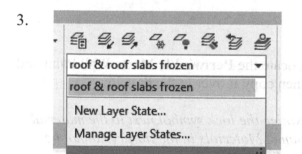   Switch to the Home ribbon.

    Set the layer state to **roof & roof slabs frozen**.

4. 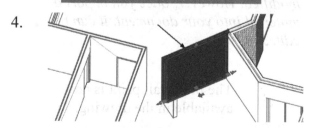   Zoom into the kitchen area. We are going to apply a different color to the kitchen wall.

    Select the wall indicated.

5. 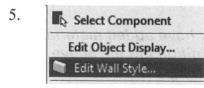   Right click and select **Edit Wall Style**.

    Notice that there are two GWB components. One GWB is for one side of the wall and one is for the other.

6.

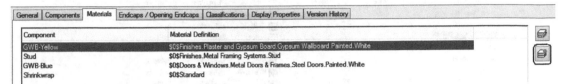

Index	Name	Priority	
1	GWB- Yellow	1210	5
2	GWB	1200	5
3	Stud	500	4
4	GWB	1200	5
5	GWB-Blue	1210	E

Rename them **GWB-Yellow** and **GWB-Blue**.

To rename, just left click in the column and edit the text.

7. Activate the Materials tab.

Component	Material Definition
GWB-Yellow	0Finishes.Plaster and Gypsum Board.Gypsum Wallboard.Painted.White
Stud	0Finishes.Metal Framing Systems.Stud
GWB-Blue	0Doors & Windows.Metal Doors & Frames.Steel Doors.Painted.White
Shrinkwrap	0Standard

Highlight the **GWB-Yellow** component.
Select the **New** button.

8.

New Name: Finishes.Wall Paint.Yellow

OK   Cancel

Type **Finishes.Wall Paint.Yellow.**

Click **OK**.

9.

Component	Material Definition
GWB-Yellow	Finishes.Wall Paint.Yellow
GWB	Finishes.Plaster and Gypsum Board.Gypsum Wallboard.Painted.White
Stud	Finishes.Metal Framing Systems.Stud
GWB	Finishes.Plaster and Gypsum Board.Gypsum Wallboard.Painted.White
GWB-Blue	Finishes.Plaster and Gypsum Board.Gypsum Wallboard.Painted.White
Shrinkwrap	Standard

Highlight the **GWB-Yellow** component.
Select the **Edit** button.

10.

Display Representations	Display Property Source	Style Override
General	Material Definition Override - Finishes.Plaste...	☑
General High Detail	Material Definition Override - Finishes.Plaste...	☑
General Low Detail	Material Definition Override - Finishes.Plaste...	☑
General Medium Det...	**Material Definition Override - Finishes....**	☑
General Presentation	Material Definition Override - Finishes.Plaste...	☑
General Reflected	Material Definition Override - Finishes.Plaste...	☑
General Reflected Scre...	Material Definition Override - Finishes.Plaste...	☑
General Screened	Material Definition Override - Finishes.Plaste...	☑

Activate the **Display Properties** tab.
Select the **Edit Display Properties** button.

11.

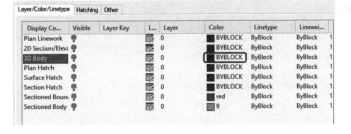

Display Co...	Visible	Layer Key	L...	Layer	Color	Linetype	Linewei...	
Plan Linework	●			0	■ BYBLOCK	ByBlock	ByBlock	1
2D Section/Eleva	●			0	■ BYBLOCK	ByBlock	ByBlock	1
3D Body	●			0	■ BYBLOCK	ByBlock	ByBlock	1
Plan Hatch	●			0	■ BYBLOCK	ByBlock	ByBlock	1
Surface Hatch	●			0	■ BYBLOCK	ByBlock	ByBlock	1
Section Hatch	●			0	■ BYBLOCK	ByBlock	ByBlock	1
Sectioned Boun	●			0	■ red	ByBlock	ByBlock	1
Sectioned Body	●			0	■ 9	ByBlock	ByBlock	1

Activate the **Layer/Color/ Linetype** tab.

Highlight **3D Body**.

Left pick on the color square.

12. 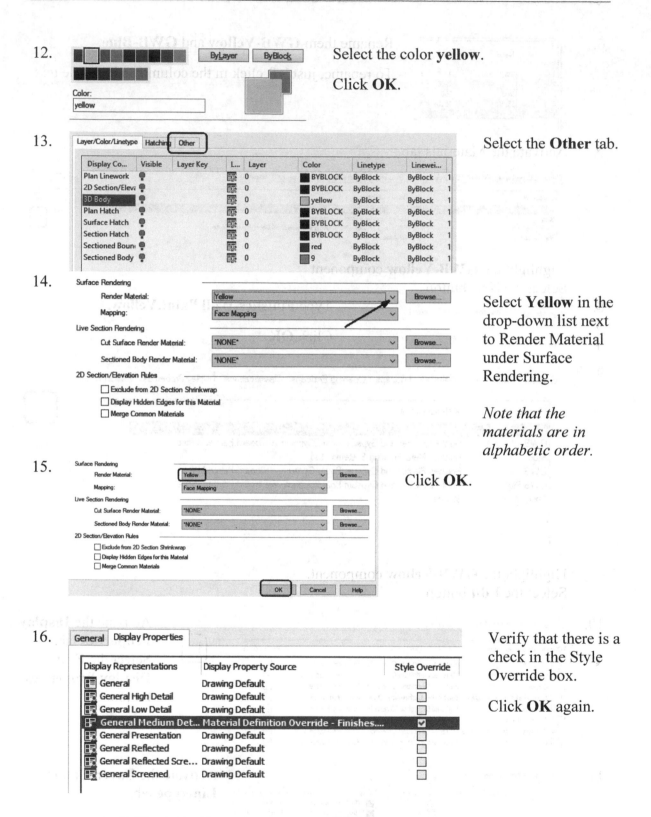 Select the color **yellow**.

Click **OK**.

13. Select the **Other** tab.

14. Select **Yellow** in the drop-down list next to Render Material under Surface Rendering.

Note that the materials are in alphabetic order.

15. Click **OK**.

16. Verify that there is a check in the Style Override box.

Click **OK** again.

17.

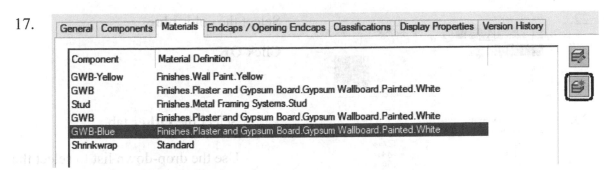

Highlight the **GWB-Blue** component.
Select the **New** button.

18.

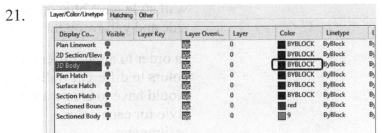

Type **Finishes.Wall Paint.Periwinkle**.

Click **OK**.

19.

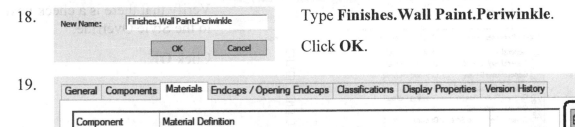

Highlight the **GWB-Blue** component.
Select the **Edit** button.

20.

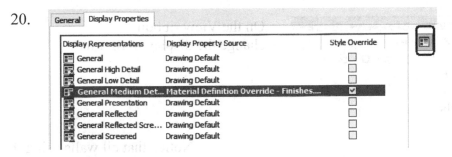

Activate the **Display Properties** tab.
Select the **Edit Display Properties** button.

21.

Activate the **Layer/Color/Linetype** tab.

Highlight **3D Body**.

Left pick on the color square.

22.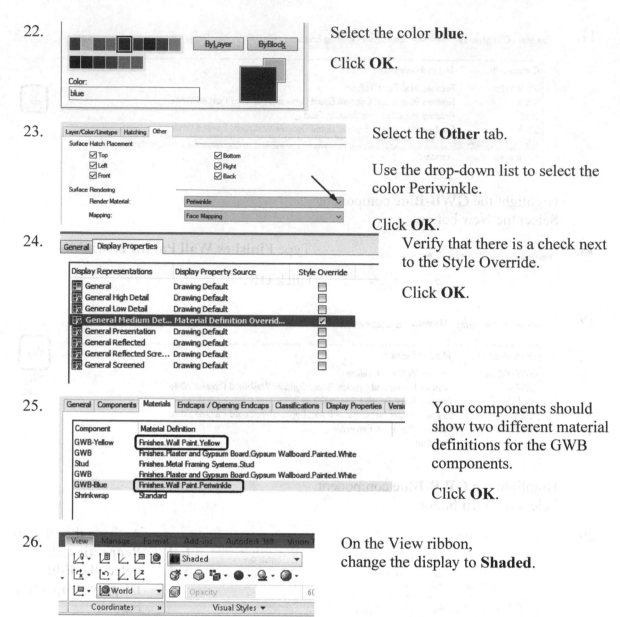

Select the color **blue**.

Click **OK**.

23. Select the **Other** tab.

Use the drop-down list to select the color Periwinkle.

Click **OK**.

24. Verify that there is a check next to the Style Override.

Click **OK**.

25. Your components should show two different material definitions for the GWB components.

Click **OK**.

26. On the View ribbon, change the display to **Shaded**.

27. Inspect the walls.

Notice that all walls using the wall style have blue on one side and yellow on the other.

In order to apply different colors to different walls, you would have to create a wall style for each color assignment.

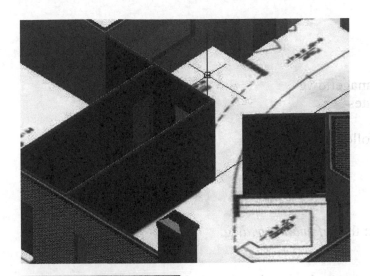

28.    To flip the color so walls are yellow/blue, simply use the direction arrows.

29. Save as *ex8-3.dwg*.

External References

Most AEC firms organize their building projects using external references. External references are files that are linked to the host drawing. Because some of the exercise files provided with this textbook use external references, you need to re-locate the external files and re-path the links.

Exercise 8-4:

Reference Manager

Drawing Name: ref_manager.dwg
Estimated Time: 5 minutes

This exercise reinforces the following skills:

- ❑ Reference Manager
- ❑ Project Navigator

1.  Open the *ref_manager.dwg*.

2. 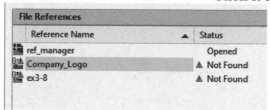 Locate the Manage Xrefs tool on the Status Bar at the bottom of the screen.
Click it to launch.

3. Highlight the *Company logo* file.

File References	
Reference Name ▲	Status
ref_manager	Opened
Company_Logo	⚠ Not Found
ex3-8	⚠ Not Found

4. In the bottom pane, locate the Saved Path field.

Date	
Found At	
Saved Path	E:\Schroff\AutoCAD 2012\exercise files\Company_Logo.jpg ...
Color System	RGB

Click on the ...
Locate the *Company_Logo.jpg* file.

5.
File name:	Company_Logo.jpg
Files of type:	All image files

Click **Open**.

6.
Would you like to apply the same path to other missing references?

c:\aca 2023 fundamentals\aca 2023 fundamentals exercises\lesson_08

Yes   No

Click **No**.

7.

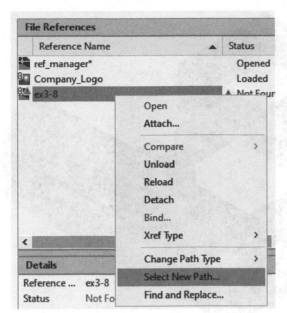

Highlight the xref drawing (the drawing name may be different) – select the external reference with the warning icon next to it.

Right click and use the drop-down to select **Select New Path**.

8.

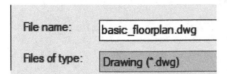

Locate and select *basic_floorplan.dwg*.
Click **Open**.

9.

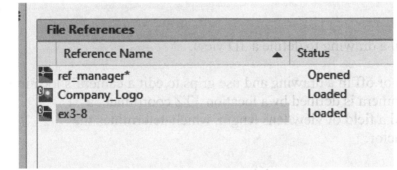

This changes the Saved Path to *basic_floorplan.dwg*.

10.

Select the **Refresh** button.

The file updates. Notice the structural members that were added in the previous lesson. The colors on the interior walls should also display properly in Shaded display mode.

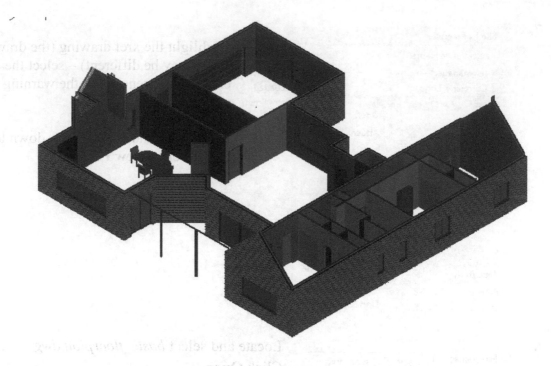

11. Save the file as *ex8-4.dwg*.

Camera

You can place a camera in a drawing to define a 3D view.

You can turn a camera on or off in a drawing and use grips to edit a camera's location, target, or lens length. A camera is defined by a location *XYZ* coordinate, a target *XYZ* coordinate, and a field of view/lens length, which determines the magnification, or zoom factor.

By default, saved cameras are named sequentially: Camera1, Camera2, and so on. You can rename a camera to better describe its view. The View Manager lists existing cameras in a drawing as well as other named views.

Exercise 8-5:

Camera View

Drawing Name: camera.dwg
Estimated Time: 20 minutes

This exercise reinforces the following skills:

- ❑ Camera
- ❑ Display Settings
- ❑ Navigation Wheel
- ❑ Saved Views
- ❑ Adjust Camera

1.   Open *camera.dwg*.

2.   Use the View Cube to orient the floor plan to a Top View with North in correct position. *You can also use the right click menu on the upper left of the display window.*

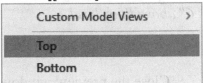

3.   Activate the Render ribbon.

On the Camera panel,

select **Create Camera**.

4.   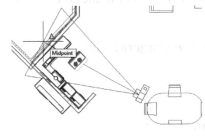 Place the Camera so it is in the dining area and looking toward the kitchen.

Click ENTER.
Select the camera so it highlights.

5.

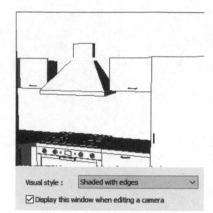

A Camera Preview window will open.

If you don't see the Camera Preview window, left click on the camera and the window should open.

Set the Visual Style to **Shaded with Edges**.

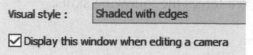

6.

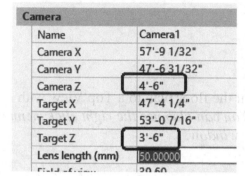

Right click on the camera and select **Properties**.

Change the Camera Z value to **3′ 6″**.

Change the Target Z to **3' 6"**.

This points the camera slightly down.

This sets the camera at 3' 6" above floor level.

7.

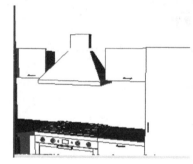

Note that the Camera Preview window updates.

Close the preview window.

Click ESC to release the camera.

8.

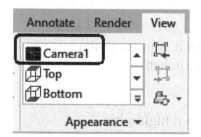

Activate the View ribbon.

Notice the Camera1 view has been added to the view lists.

Left click to activate the Camera1 view.

9.

Switch to the Render ribbon.

Click **Adjust**.

10.

You can use the dialog to adjust the view to your liking.

Click **OK**.
Switch to the View ribbon.

11.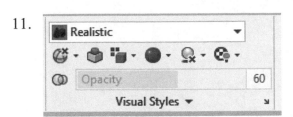

Change the Display Style to **Realistic**.

12.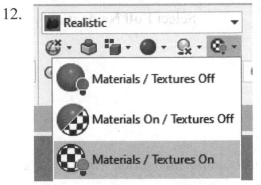

Set Materials/Textures **On**.

13.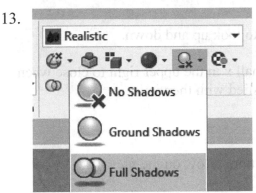

Set **Full Shadows** on.

14.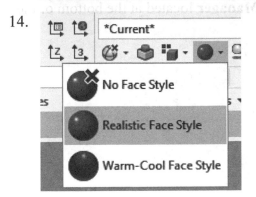

Switch to the View ribbon.

Set **Realistic Face Style** on.

15.

The Camera view should update as the different settings are applied.

16.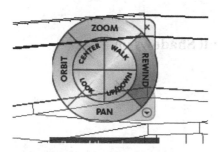

Select **Full Navigation**.

17.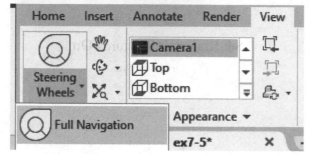

Use the controls on the navigation wheel to adjust the view.

Try LOOK to look up and down.

Click the small x in the upper right to close when you are satisfied with the view.

18.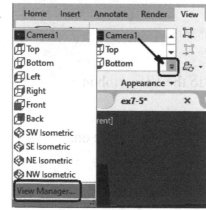

Click the bottom arrow on the view list.

Select **View Manager** located at the bottom of the view list.

19.

Select **New**.

20.

Name the view **Kitchen.**

Select the down arrow to expand the dialog.

21.

Set the Visual style to **Realistic**.

Set the background to Solid.

22.

Click in the Color box to set the color.

23.

Set the background color to 255,255,255 which is white.

Click **OK**.

Close all the dialogs other than the View Manager.

24.

Current View: Current

Views

- Current
- Model Views
  - Camera1
  - Kitchen
- Layout Views
- Preset Views

The saved view is now listed under Model Views.

Click **OK**.

Save as *ex8-5.dwg*.

Rendering

Rendering is the process where you create photo-realistic images of your model. It can take a lot of trial and error to set up the lighting and materials as well as the best camera angle before you get an image you like. Be patient. The more you practice, the easier it gets. Make small incremental changes – like adding more or less lighting – turning shadows off and on – changing materials – and render between each change so you can understand which changes make the images better and which changes make the images worse.

If you have created a rendering that is close to what you want, you may be able to use the built-in image editor to make it lighter/darker or add more contrast. That may be enough to get the job over the finish line.

Exercise 8-6:
Create Rendering

Drawing Name: rendering.dwg
Estimated Time: 5 minutes

This exercise reinforces the following skills:

- ❑ Render
- ❑ Display Settings
- ❑ Set Location

You need an internet connection in order to perform this exercise.

1. Open *rendering.dwg*.

2. Set the Top view active.

3. Activate the Render ribbon.

4.  Select **Sky Background and Illumination**.

5.

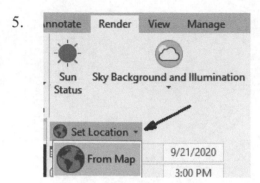

In the drop-down panel under Sky Background and Illumination, select

Set Location→From Map.

6.

Do you want to use Online Map Data?

Online Map Data enables you to use an online service to display maps in AutoCAD. Please sign into your Autodesk account to access online maps.

By accessing or using this service, you understand and agree that you will be subject to, have read and agree to be bound by the terms of use and privacy policies referenced therein: Online Map Data - Terms of Service.

☐ Remember my choice   Yes   No

Click **Yes** if you have internet access.

7.

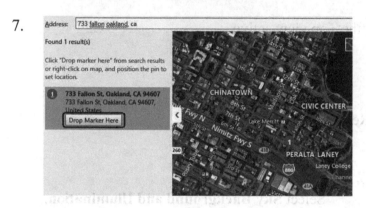

Type in an address in the search field where your project is located.

Once it is located on the map, select **Drop Marker here.**

Click **Next**.

8.

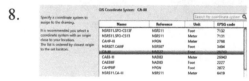

Select the coordinate system to be applied to the project.

Highlight **CA-III**, then Click **Next**.

9.

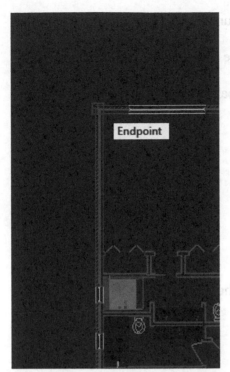

Select the upper left corner of the building as the site location.

Click ENTER to accept the default North location.

Your building model will appear on a satellite image of the address you provided.

This is useful to provide an idea of how the model will look in the area where it will be constructed.

10.

Switch back to the Render ribbon.

Select the small down arrow at the bottom right of the Sun & Location panel.

11.

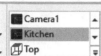

Scroll to Sun Angle Calculator.

Set the time to **3:00 PM**.

Close the palette.

Sun Angle Calculator	▼
Date	9/21/2021
Time	3:00 PM
Daylight S...	No
Azimuth	238.71
Altitude	34.42
Source Ve...	-23/32",-7/16",...

12.

Activate the View ribbon.

Set the view to **Kitchen**.

13.

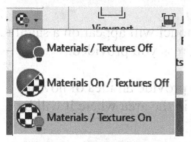

Full Shadows should be set to **ON**.

14.

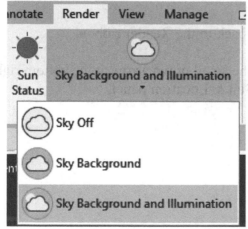

Materials and Textures should be set to **ON**.

15.

Switch to the Render ribbon.

Set the **Sky Background and Illumination on**.

You should see a slight change in the view.

16.

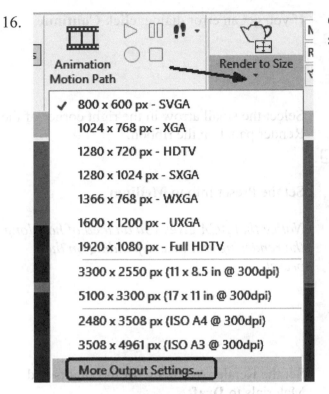

On the Render to Size drop-down, select the **More Output Settings**.

17.

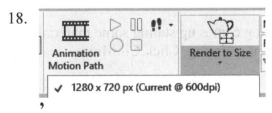

Set the Image Size to **1280x720 px**.
Set the Resolution to **600**.
Click **Save**.

18.

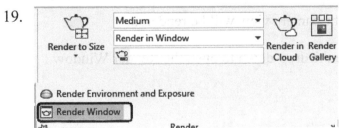

Click on **Render to Size**.

Note the setting you created is enabled.

19.

Expand the Render panel in the ribbon.

Select **Render Window**.

This opens a window where you will see the rendering.

20. If you get an error dialog, click **Continue**.

21. Select the small arrow in the right corner of the Render panel on the ribbon.

22. Set the Preset Info to **Medium**.

   Notice that ACA gives you an idea of how long the rendering will take depending on the preset.

23. Set the Render Accuracy for the Lights and Materials to **Draft.**

24. Click on the **Render** button.

25. If a dialog comes up stating some material images are missing, Click **Continue**.

26. A window will come up where the camera view will be rendered.

   If you don't see the window, use the drop-down to open the Render Window.

27. 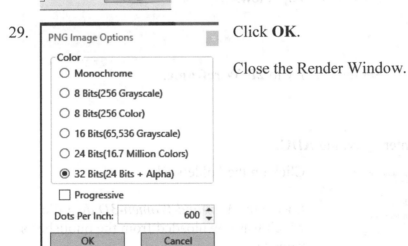 Select the **Save** icon to save the rendered image.

28. | | |
    |---|---|
    | File name: | kitchen |
    | Files of type: | PNG(*.png) |

    Save the image file to the exercise folder.

29. **PNG Image Options**

    Color
    - ○ Monochrome
    - ○ 8 Bits(256 Grayscale)
    - ○ 8 Bits(256 Color)
    - ○ 16 Bits(65,536 Grayscale)
    - ○ 24 Bits(16.7 Million Colors)
    - ◉ 32 Bits(24 Bits + Alpha)

    ☐ Progressive

    Dots Per Inch: 600

    [OK]   [Cancel]

    Click **OK**.

    Close the Render Window.

30. Save as *ex8-6.dwg*.

Exercise 8-7:
Render in Cloud

Drawing Name: cloud_render.dwg
Estimated Time: 5 minutes

This exercise reinforces the following skills:

- ❑ Render in Cloud
- ❑ Design Center
- ❑ Blocks

Autodesk offers a service where you can render your model on their server using an internet connection. You need to register for an account. Educators and students can create a free account. The advantage of using the Cloud for rendering is that you can continue to work while the rendering is being processed. You do need an internet connection for this to work.

1.   Open *cloud_render.dwg*.

2.   Activate the View ribbon.

         Switch to a **Top** View.

3.   Set the Display to **2D Wireframe**.

4.   Launch the Design Center by typing **ADC**.

5.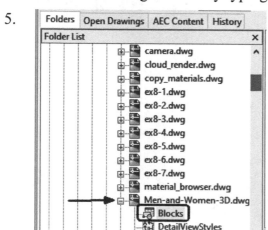
     Click on the Folders tab.

     Locate the *Men-and-Women-3D.dwg* file which was downloaded from the publisher's website.

     Click on **Blocks.**

6. 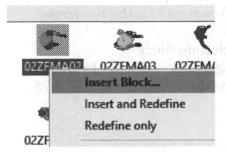 Select one of the blocks.

Right click and select **Insert Block**.

I selected 02ZFMA02.

7. 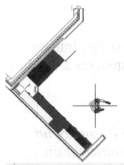 Set the Scale to **1.0**.
Enable **Uniform Scale**.

Enable **Insertion Point**.

Enable **Rotation**.

Click **OK**.

8.  Place the block inside the kitchen area.

Close the Design Center.

9. 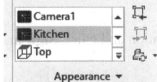 Switch to the **Kitchen** view.

10.  *You will see the 3D figure inside the kitchen.*

Use the Full Navigation Wheel to adjust the view as needed.

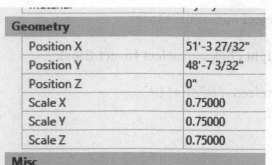

If you want to resize the person:

Select the block.
Right click and select **Properties**.
Change the Scale values to **0.75**.

The block will adjust scale accordingly.

In order to render using Autodesk's Cloud Service, you need to sign into an Autodesk account. Registration is free and storage is free to students and educators.

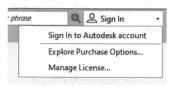

Sign in to your Autodesk account.

You can create an account for free if you are an educator, student or if you are on subscription.

11.

Enter your log-in information and Click **Sign in**.

12.

Your user name should display once you are signed in.

13. Save the file as *ex8-7.dwg*.

You must save the file before you can render.

14. Activate the Render ribbon.

Select **Render in Cloud**.

15. Your free preview renderings will appear in your Autodesk Rendering Account online. From there you can view, download, and share, or fine tune render settings to create higher resolution final images, interactive panoramas, and more.

☑ Notify me by e-mail when complete.

Set the Model View to render the current view.

Click the **Start Rendering** model.

16.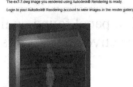

A small bubble message will appear when the rendering is completed.

17. **AUTODESK.** RENDERING

Dear Elise Moss!

The ex7-7.dwg image you rendered using Autodesk® Rendering is ready.

Login to your Autodesk® Rendering account to view images in the render gallery.

You will also receive an email when the rendering is complete. The email is sent to the address you used to sign into your account.

18. Render Gallery

Select **Render Gallery** on the Render ribbon to see your renderings.

19. MY RENDERINGS

Click on the My Renderings tab on the browser to review your renderings.

20. You can also select View completed rendering from the drop-down menu under your sign in.

Sign Out

Account Details

Explore Purchase Options...

Manage License...

Privacy Settings...

View completed rendering

You may need to sign into the gallery in order to see your rendering.

Compare the quality of the rendering performed using the Cloud with the rendering done in the previous exercise.

I have found that rendering using the Cloud is faster and produces better quality images. It also allows me to continue working while the rendering is being processed.

Walk-Through

You can create 3D preview animations and adjust the settings before you create a motion path animation.

Preview animations are created with the controls on the Animation panel found on the ribbon and the 3D navigation tools. Once a 3D navigation tool is active, the controls on the Animation panel are enabled to start recording an animation.

Walk-through animations are useful when reviewing building sites or discussing building renovations.

The easiest way to create a walk-through is to sketch out the path using a polyline. Then, attach a camera to follow the desired path.

Exercise 8-8:
Create an Animation

Drawing Name: animation.dwg
Estimated Time: 5 minutes

This exercise reinforces the following skills:

❑   Walk-Through

It will take the software several minutes to compile the animation. It is a good idea to do this exercise right before a 30-minute break to allow it time to create the animation during the break.

1.   Open *animation.dwg*.

2.   Activate the View ribbon.

        Switch to a **Top** View.

3.   Set the view display to **2D Wireframe**.

4.   Activate the Home ribbon.

        Set the layer named **Walkthrough** Current.

        This layer is set so it doesn't plot.

5.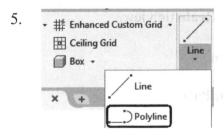
        Select the **Polyline** tool located under the Line drop-down list.

6.

Draw a polyline that travels through the floor plan to view different rooms.

7.

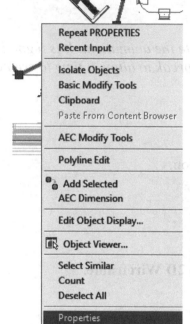

Select the polyline.

Right click and select **Properties**.

8.

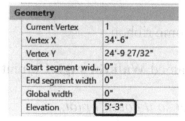

Set the elevation of the polyline to **5' 3"**.

If you don't raise the line, your camera path is going to be along the floor.

Click **ESC** to release the selected polyline.

9.   [−][SW Isometric][Realistic]

Switch to a **SW Isometric Realistic** view.

10.

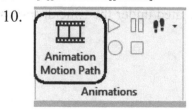

Activate the Render ribbon.

Select **Animation Motion Path**.

11.

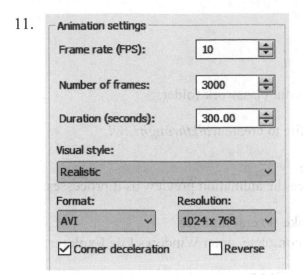

Set the Frame Rate to **10**.

Set the Number of frames to **3000**.

Set the Duration to **300**.

Set the Visual Style to **Realistic**.

Set the Format to **AVI**.
This will allow you to play the animation in Windows Media Player.

Set the Resolution to **1024 x 768**.

12.

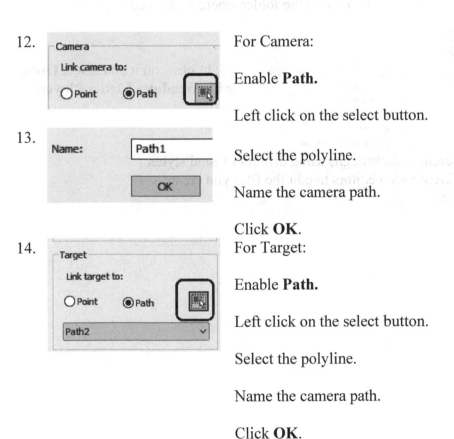

For Camera:

Enable **Path.**

Left click on the select button.

13.

Select the polyline.

Name the camera path.

Click **OK**.

14.

For Target:

Enable **Path.**

Left click on the select button.

Select the polyline.

Name the camera path.

Click **OK**.

15.
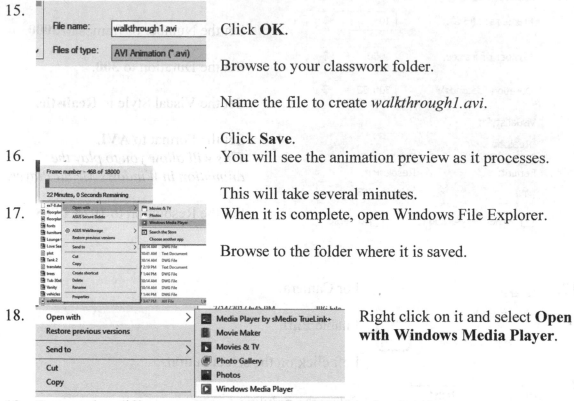

Click **OK**.

Browse to your classwork folder.

Name the file to create *walkthrough1.avi*.

Click **Save**.

16. You will see the animation preview as it processes.

This will take several minutes.

17. When it is complete, open Windows File Explorer.

Browse to the folder where it is saved.

18. Right click on it and select **Open with Windows Media Player**.

19. Try creating different walkthroughs using different visual styles.
You can use different video editors to edit the files you create.

20. Save as *ex8-8.dwg*.

Exercise 8-9:
Using the Sun & Sky

Drawing Name: sun_study.dwg
Estimated Time: 15 minutes

This exercise reinforces the following skills:

- ❏ Create a saved view
- ❏ Apply sunlight
- ❏ Add an image as background
- ❏ Adjust background

1.   Open *sun_study.dwg*.

2.   Right click on the ViewCube and select **Parallel**.

How does the view shift?

3.   Right click on the ViewCube and select **Perspective**.

How does the view shift?

4.   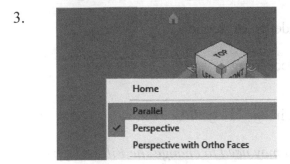  Double click the wheel on the mouse to Zoom Extents.

Adjust the view so that the back edge of the slab is horizontal in the view.

Hint: Switch to a Front view then tip the view slightly forward.

5.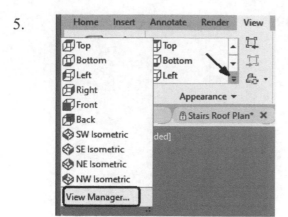

Switch to the View tab.

Launch the **View Manager**.

6.

Click **New**.

7.

Type **Front Isometric** for the View Name.

8.

Expand the lower pane of the dialog.

Under Background:

Select **Image**.

9.

Set image to *sky.JPG*.

You may need to change the Files of type to JFIF. This file is included in the downloaded exercise files.

Click **Open**.

10.

Click **OK**.

11.

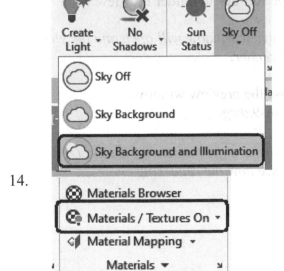

Click **OK**.

Set the Front Isometric as the Current View.

You have now saved the view.

Close the View Manager.

12.

Type **Render** to render the view.

A preview window will open and the view will render.

Close the preview window.

13.

Switch to the Render tab.

Enable **Sky Background and Illumination**.

14.

Enable **Materials/Textures On**.

15.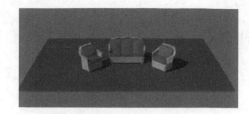

Type **Render** to render the view.

A preview window will open and the view will render.

Close the preview window.

16.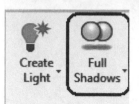

Enable **Full Shadows**.

17.

Click the small arrow in the lower right corner of the Sun & Location panel.

18.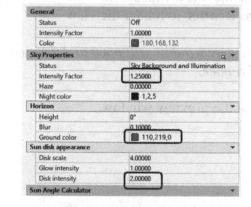

Set the Intensity Factor to **1.25**.
Set the Ground color to **110,219,0**.
(This is a light green.)
Set the Disk Intensity to **2**.

Close the dialog.

19.

Type **Render** to render the view.

A preview window will open and the view will render.

Close the preview window.

20. Save as *ex8-9.dwg*.

This exercise let you try different settings and compare the rendering results. Try applying different backgrounds and lighting to see how the rendering changes.

Exercise 8-10:
Applying Materials

Drawing Name: materials.dwg
Estimated Time: 45 minutes

This exercise reinforces the following skills:

- Navigate the Materials Browser
- Apply materials to model components
- Edit Styles
- Change environment
- View Manager
- Render

1.   Open *materials.dwg*.

2.   [−][SW Isometric][Realistic]   Set the display to **Realistic**.

   If you don't set the display to Realistic, you won't see the material changes.

3.   Switch to the Render ribbon.

   Click on **Materials Browser**.

4.   Locate **Beechwood – Honey** and load into the drawing.

   This is for the flooring.

5.   Locate **Glass Clear - Light** and load into the drawing.

   This is for the window glazing.

6.   Locate **Wall Texture - Salmon** and load into the drawing.

   This is for the walls.

7.

Locate **Sand Wall Paint** and load into the drawing.

This is for the wall and window trim.

8.

Change the Material Browser View to use a Thumbnail View.

Now you can see swatches of all the materials available in your drawing.

Close the Materials Browser.

9.

Select a wall.

On the ribbon:
Select **Edit Style**.

10.

Click the Materials tab.

Highlight the top component for the wall.
Click on the **Edit Material** button.

11.

Click the Display Properties tab.

Click on the override button.

12.

Click the Other tab.

Set the Render Material to **Wall Texture – Salmon**.

Click **OK**.

Close all the dialogs.

Click ESC to release the selection.

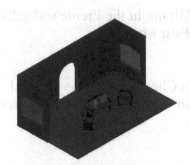

You should see the new material on the walls.

If you don't see the material change, verify that your display is set to **Realistic**.

This only changes the wall style used in this drawing.

13. Click on a window.
On the ribbon:
Select **Edit Style**.

14. Click the Materials tab.

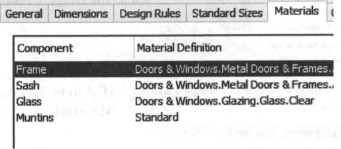

Highlight the **Frame Component**.
Click on **New Material**.

15. New Name: Sand Paint OK

Create a new material called **sand paint**.

Click **OK**.

16. | Component | Material Definition |
| --- | --- |
| Frame | Sand Paint |
| Sash | Doors & Windows.Metal Doors & Frames. |
| Glass | Doors & Windows.Glazing.Glass.Clear |
| Muntins | Standard |

The Frame is now assigned the new material.

17. Highlight the Glass Component.
Click on **New Material**.

18. New Name: Glass - Clear Light OK

Create a new material called **Glass – Clear Light**.

Click **OK**.

19. | Component | Material Definition |
| --- | --- |
| Frame | Sand Paint |
| Sash | Doors & Windows.Metal |
| Glass | Glass - Clear Light |
| Muntins | Standard |

The Glass is now assigned the new material.

20. 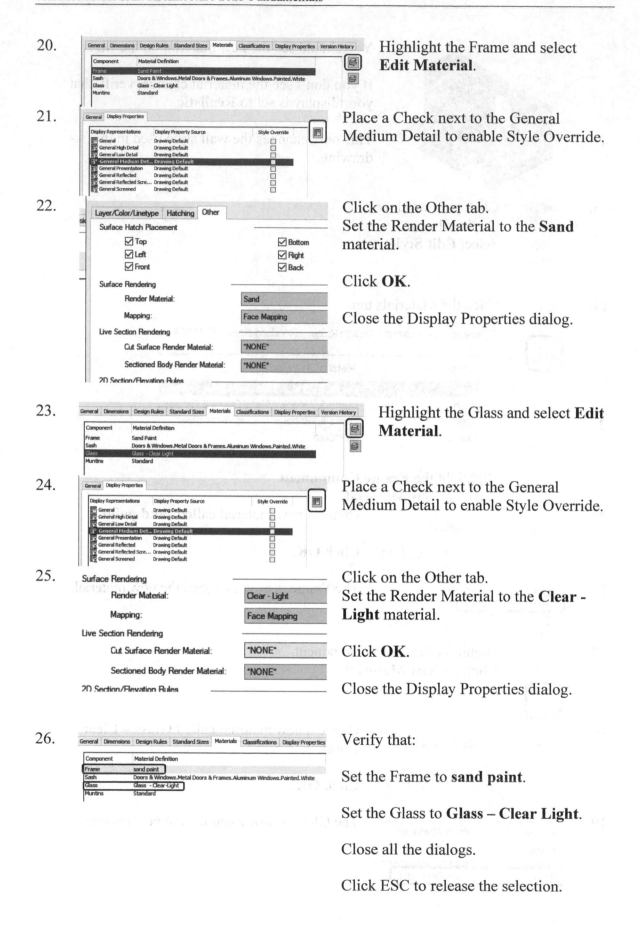 Highlight the Frame and select **Edit Material**.

21. Place a Check next to the General Medium Detail to enable Style Override.

22. Click on the Other tab.
Set the Render Material to the **Sand** material.

Click **OK**.

Close the Display Properties dialog.

23. Highlight the Glass and select **Edit Material**.

24. Place a Check next to the General Medium Detail to enable Style Override.

25. Click on the Other tab.
Set the Render Material to the **Clear - Light** material.

Click **OK**.

Close the Display Properties dialog.

26. Verify that:

Set the Frame to **sand paint**.

Set the Glass to **Glass – Clear Light**.

Close all the dialogs.

Click ESC to release the selection.

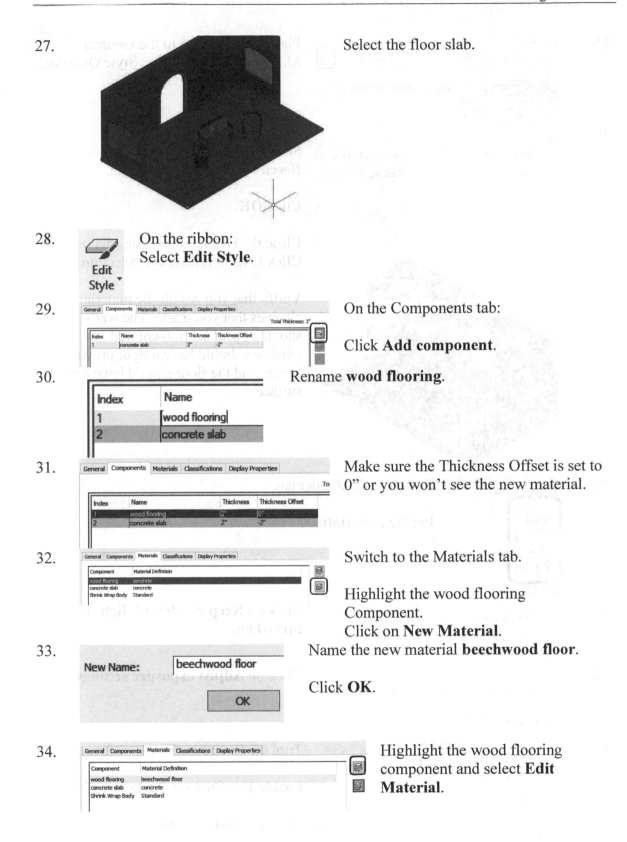

27.   Select the floor slab.

28. On the ribbon:
Select **Edit Style**.

29. On the Components tab:

Click **Add component**.

30. Rename **wood flooring**.

31. Make sure the Thickness Offset is set to 0" or you won't see the new material.

32. Switch to the Materials tab.

Highlight the wood flooring Component.
Click on **New Material**.

33. Name the new material **beechwood floor**.

Click **OK**.

34. Highlight the wood flooring component and select **Edit Material**.

35.

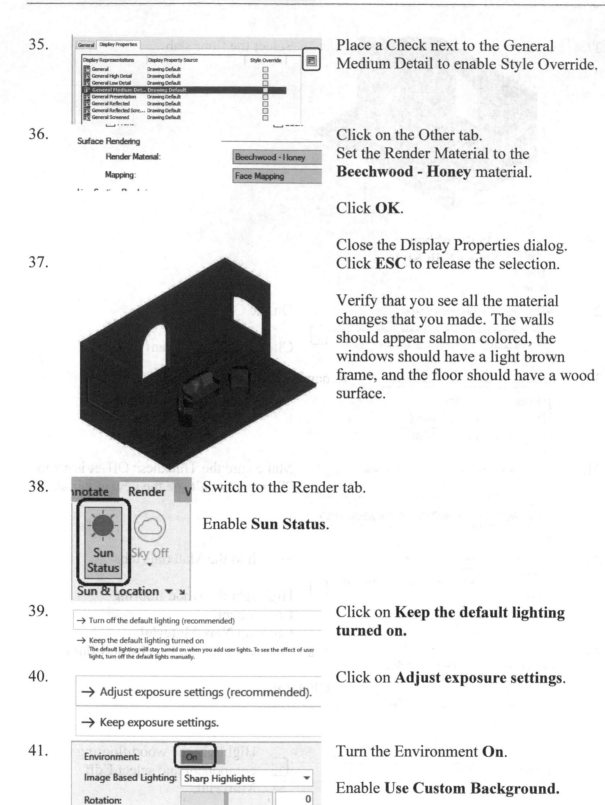

Place a Check next to the General Medium Detail to enable Style Override.

36.

Click on the Other tab.
Set the Render Material to the **Beechwood - Honey** material.

Click **OK**.

37.

Close the Display Properties dialog.
Click **ESC** to release the selection.

Verify that you see all the material changes that you made. The walls should appear salmon colored, the windows should have a light brown frame, and the floor should have a wood surface.

38.

Switch to the Render tab.

Enable **Sun Status**.

39.

Click on **Keep the default lighting turned on.**

40.

Click on **Adjust exposure settings**.

41.

Turn the Environment **On**.

Enable **Use Custom Background.**

Click on **Background…**

42.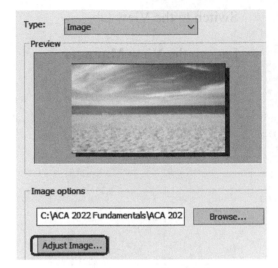

Set Type to **Image**.

Click **Browse**.

Select the *Beach.jpg* file in the downloaded exercise files.

Click **Open**.

Click on **Adjust Image**.

43.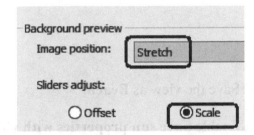

Select **Stretch** for Image position.

Enable **Scale**.

Click **OK**.

Close the dialogs.

44.

Orbit, pan and zoom to position the model on the beach so it appears proper.

45.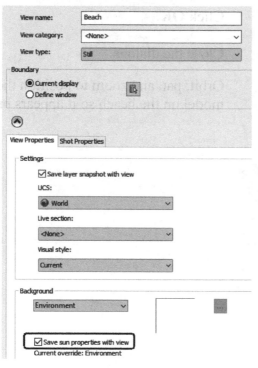

Switch to the View tab.

Launch the **View Manager**.

46. New...

Click **New**.

47.

Save the view as **Beach**.

Enable **Save sun properties with view**.

Click **OK**.

48.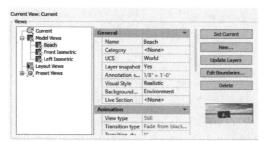

Click **OK**.

49.  Type **RENDER**.

The preview window will appear and the view will render.

Save as *ex8-10.dwg*.

Exercise 8-11:
Applying Materials to Furniture

Drawing Name: Wicker chair and sofa.dwg
Estimated Time: 15 minutes

This exercise reinforces the following skills:

- ❏ Navigate the Materials Browser
- ❏ Apply materials to model components
- ❏ Customize materials

1.

| File name: | Wicker chair and sofa.dwg |
| Files of type: | Drawing (*.dwg) |

Open the *Wicker chair and sofa* file in the exercises folder.

2.

| File name: | Wicker chair and sofa - Sea Fabric.dwg |
| Files of type: | AutoCAD 2018 Drawing (*.dwg) |

Save the file with a new name - *Wicker chair and sofa – Sea Fabric.*

3.

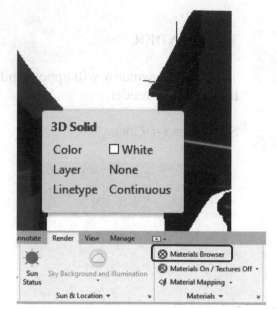

Use explode to explode the blocks until you see solids when you hover over the elements.

Type **EXPLODE, ALL, ENTER**.

Repeat until you see 3D solid when you hover over each item.

4.

Switch to the Render ribbon.

Click on **Materials Browser**.

5.

Pebbled - Black

Pebbled - Light Brown

Pebbled - Mauve

Locate **Pebbled – Mauve** and load into the drawing.

This is for the chair fabric.

6.

Chrome - Polished Brushed	Autod...	Metal	Metal	
Chrome - Polished Brushed Heavy	Autod...	Metal	Metal	
Chrome - Satin	Autod...	Metal	Metal	
Chrome - Satin 1	Autod...	Generic	Metal	

Locate **Chrome - Satin** and load into the drawing.

This is for the chair frame.

7.

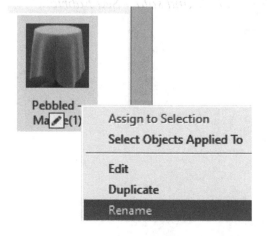

Highlight the Pebbled – Mauve fabric.

Right click and select **Duplicate**.

8.

Right click on the duplicate material.

Select **Rename**.

9.  Rename **Sea Fabric**.

10. 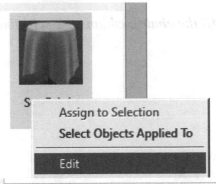 Select the Sea Fabric material.

Right click and select **Edit**.

11.  Click on the image name and set to *sea fabric.jpg*.
This file is available as part of the exercise downloads.

Change the Color to **255,255,255**.
This is white.

Close the Materials Editor.

12. 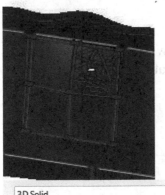 Select one of the cushions.

Right click and select **Properties**.

13. 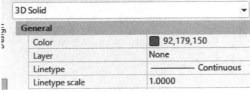 On the Properties palette:

Note the color is set to **92, 179,150**.

Click **ESC** to release the selection.

14.

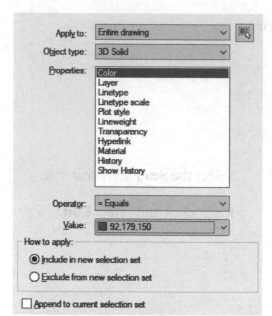

Type **QSELECT** for quickselect.

Set the Object type to **3D Solid**.

Set the Color Equals to **92, 179, 150**.

Click **OK**.

All the chair cushions are selected.

15.

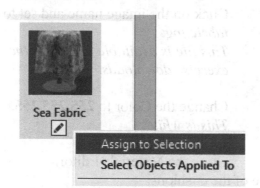

Highlight the Sea Fabric material in the Materials Browser.

Right click and select **Assign to Selection**.

Click ESC to release the selection.

[Realistic]

If the view is set to Realistic, you should see the material as it is applied.

16.

Select one of the chair frame elements.

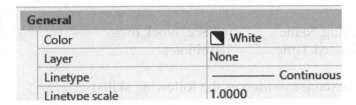

General	
Color	White
Layer	None
Linetype	—————— Continuous
Linetype scale	1.0000

The Color is set to White.

Click ESC to release the selection.

17.

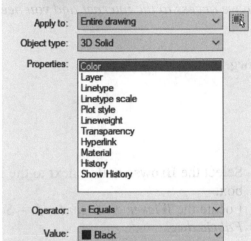

Type **QSELECT** for quickselect.

Set the Object type to **3D Solid**.

Set the Color Equals to **White**.

Click **OK**.

The chair frames are selected.

18.

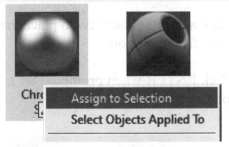

Highlight the **Chrome - Satin** material.

Right click and select **Assign to Selection**.

Click **ESC** to release the selection.

19.

The drawing has been redefined with new materials.

Save as *Wicker chair and sofa – Sea Fabric.dwg*

Exercise 8-12:
Replace a Block

Drawing Name: replace_block.dwg
Estimated Time: 10 minutes

This exercise reinforces the following skills:

- ❑ Sync Blocks
- ❑ Replace a block
- ❑ Render

In order to sync blocks, you need to have access to the internet and you need to be signed in to your Autodesk account.

1.   Open *replace_block.dwg.*

2. Type **INSERT**.

   Click ENTER.

3.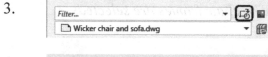

   Select the **Browse** button next to the Filter box.

4.   File name: Wicker chair and sofa - Sea Fabric.dwg

   Files of type: Drawing (*.dwg)

   Locate the *Wicker chair and sofa – Sea Fabric.dwg.*

   Click **Open**.

   Click a point in the view to place the block away from the room.

   Click ENTER when prompted for a rotation angle.

5.   (i) Blocks are only available across devices if ✕ you select cloud storage.

   Block Sync Settings...

   This dialog will appear.

   You can now save your blocks in your Autodesk account on the cloud, so you can locate them from any device.

   Click **Block Sync Settings**.

6.

Recent and Favorite Block Libraries:

C:\ACA 2023 Fundamentals\ACA 2023 Fundamentals exercises [...]

Blocks are only accessible across devices if you use cloud storage.

Storage Permission

☑ Remember my cloud storage user names across devices.

Learn more

OK   Cancel

Select the folder where you are storing your exercise files.

Enable **Storage Permission**.

Click **OK**.

In a work environment, you might store all your blocks in a folder and select that folder. To use cloud storage, select your Dropbox folder or similar cloud folder.

Close the dialog.
Type **BLOCKREPLACE**.

Scroll down and select *Wicker chair and sofa*.

Click **OK**.

7.

Select the block to be replaced

Group-2-1
Group-3-1
Group-4-1
Group-5-1
Group-6-1
Group-7-1
Group-8-1
Group-9-1
Wicker chair and sofa

Pick<   Wicker chair and sofa

OK   Cancel   Help

8.

Select a block to replace Wicker chair and sofa

Group-3-1
Group-4-1
Group-5-1
Group-6-1
Group-7-1
Group-8-1
Group-9-1
Wicker chair and sofa
Wicker chair and sofa - Sea Fabric

Pick<   Wicker chair and sofa - Sea Fabric

OK   Cancel   Help

Scroll down and select the *Wicker chair and sofa – Sea Fabric* block.

Click **OK**.

9.

▼Purge unreferenced items when finished? <Y>:

Click ENTER.

10.

The block is replaced using the same location and orientation.

Delete the block that was inserted and isn't needed any more.

Note that you had to fully insert the new block into the drawing before you could replace the block.

11.

Type RENDER.

The preview window will appear and the view will render.

Save as *ex8-12.dwg*.

Exercise 8-13:
Using Artificial Light

Drawing Name: lights.dwg
Estimated Time: 15 minutes

This exercise reinforces the following skills:

- ❑ Add lights to a model
- ❑ Edit light properties

1. Open *lights.dwg*.

2.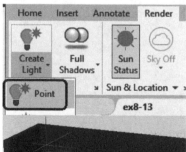

Switch to the Render ribbon.

Select the **Point** light tool.

3.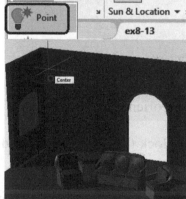

Select the center of the circle on the left as the location for the point light.

Click ENTER to accept all the default settings for the light.

If you have difficulty placing the lights, freeze all the layers except for Light Locator.

4.

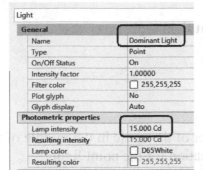

Select the point light.
Right click and select **Properties**.

Change the Name to **Dominant Light**.

Change the Lamp intensity to **15.00 Cd**.

Did you notice that the size of the light glyph updates?

Click ESC to release the selection.

5.

Enable **Full Shadows**.
Enable **Sky Background and Illumination**.

6.

→ Adjust exposure settings (recommended).

→ Keep exposure settings.

Click on **Sun Status**.
Click on **Adjust exposure settings**.

7.

Set Environment to **On**.

Set the Image Based Lighting to **Warm Light**.

Enable **Use IBL Image as Background**.

Close the palette.

8.

Render to Size

Click **Render to Size** on the ribbon.

The Render Window will appear.

You will see a new rendering.

Close the Render Window.

9.

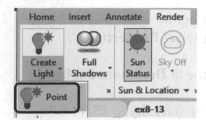

Select the **Point** light tool.

10.

Select the center of the circle in the middle of the room as the location for the point light.

Click ENTER to accept all the default settings for the light.

If you have difficulty placing the lights, freeze all the layers except for Light Locator.

11.

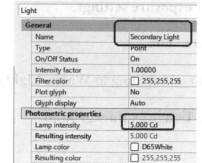

Select the point light.
Right click and select **Properties**.

Change the Name to **Secondary Light**.

Change the Lamp intensity to **5.00 Cd**.

Click ESC to release the selection.

12.

Switch to the Home ribbon.

Use **Thaw All** to thaw any frozen layers.

13.

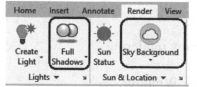

Switch to the Render ribbon.

Enable **Full Shadows**.
Enable **Sky Background**.

14.

Click on **Sun Status**.

15.

→ Adjust exposure settings (recommended).

→ Keep exposure settings.

Click on **Adjust exposure settings**.

16.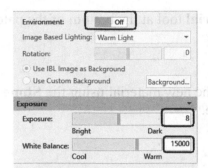

Set Environment to **Off**.

Set the Exposure to **8**.

Set the White Balance to **15000.**

Close the palette.

17.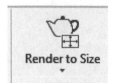

Click **Render to Size** on the ribbon.

The Render Window will appear.

You will see a new rendering.

Close the Render Window.

Save as *ex8-13.dwg*.

Exercise 8-14:
Mapping Materials

Drawing Name: medallion.dwg
Estimated Time: 15 minutes

This exercise reinforces the following skills:

- ❑ Create a material
- ❑ Apply material to a face.

1. Open *medallion.dwg*.

2.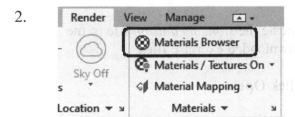

Switch to the Render tab.

Click on Materials Browser.

3. Select the **Create Material** tool at the bottom of the palette.

4. 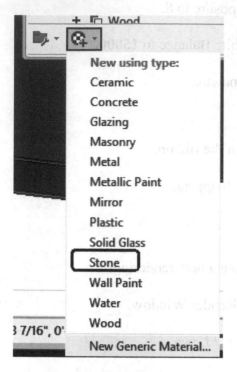 Create the new material using the **Stone** template.

5. 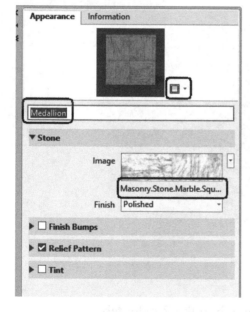 Change the name to Medallion.

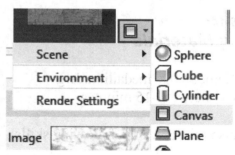

Change the Scene to **Canvas**.

Click on the image name.

6. File name: stone medallion.JPG

Locate the *stone medallion* file in the downloaded exercise files.

Click **Open**.

7.

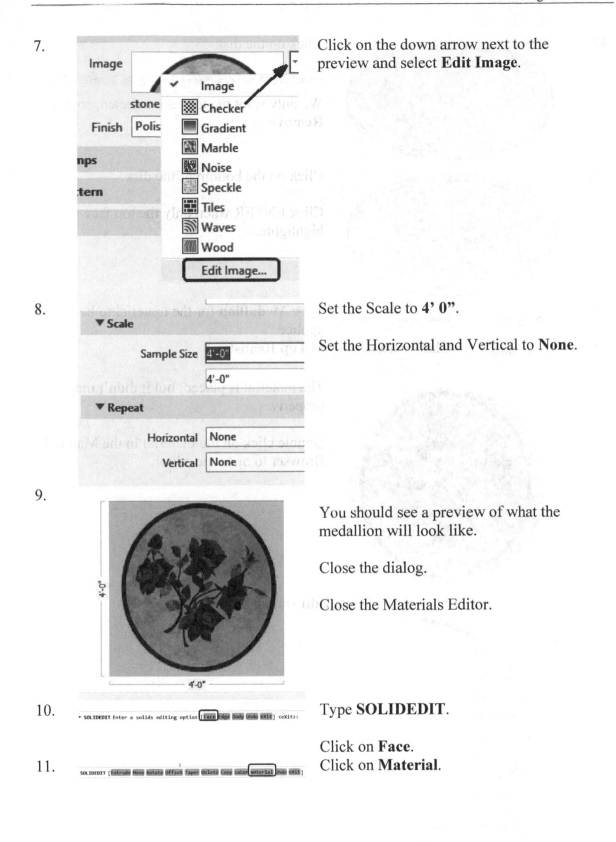

Click on the down arrow next to the preview and select **Edit Image**.

8. Set the Scale to **4' 0"**.

Set the Horizontal and Vertical to **None**.

9. You should see a preview of what the medallion will look like.

Close the dialog.

Close the Materials Editor.

10. Type **SOLIDEDIT**.

SOLIDEDIT Enter a solids editing option [Face Edge Body Undo eXit] <eXit>:

Click on **Face**.

11. Click on **Material**.

SOLIDEDIT [Extrude Move Rotate Offset Taper Delete Copy coLor mAterial Undo eXit]

12. 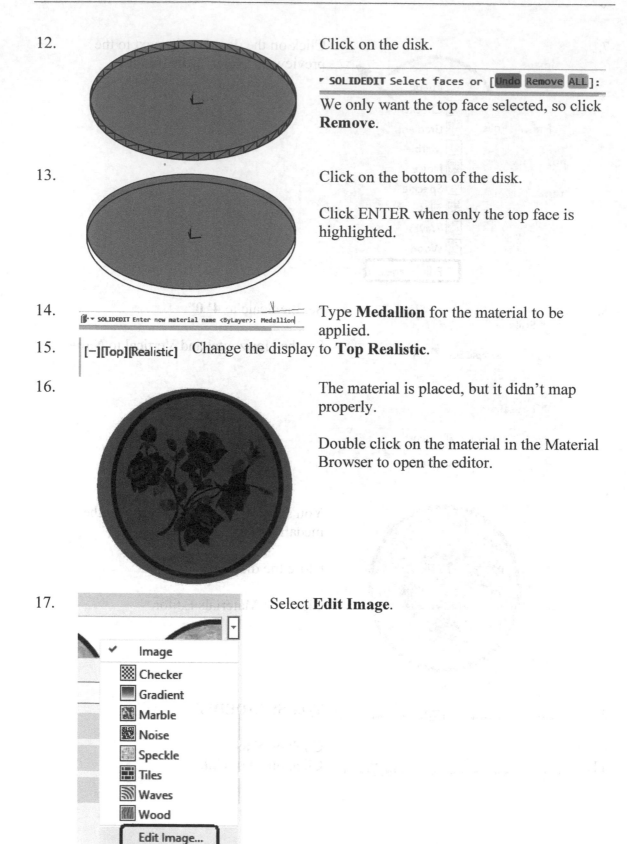 Click on the disk.

SOLIDEDIT Select faces or [Undo Remove ALL]:

We only want the top face selected, so click **Remove**.

13. Click on the bottom of the disk.

Click ENTER when only the top face is highlighted.

14. **SOLIDEDIT** Enter new material name <ByLayer>: Medallion

Type **Medallion** for the material to be applied.

15. [−][Top][Realistic]   Change the display to **Top Realistic**.

16. The material is placed, but it didn't map properly.

Double click on the material in the Material Browser to open the editor.

17. Select **Edit Image**.

✓ Image
▧ Checker
▦ Gradient
▨ Marble
▦ Noise
▩ Speckle
▤ Tiles
▨ Waves
▥ Wood

Edit Image...

18.

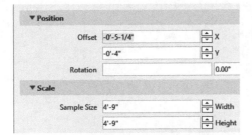

Adjust the Position and the Scale until the preview looks correct.

Close the dialogs.

19.

Save as *flower medallion.dwg*.

Extra: *Place the medallion in the beach scene you did in the previous exercise and see if you can create a nice rendering.*

Notes:

QUIZ 8

True or False

1. When rendering, it is useful to save named views using the View Manager.

2. Sky illumination can be used to add extra light to a scene.

3. You can use an image or a color to change the background of a scene.

4. To change the material used in a wall, you need to modify the wall style.

5. When mapping a material, you can only use grips to adjust the placement.

Multiple Choice

6. Materials can be displayed in the following views: (select all that apply)

    A. plans
    B. sections
    C. elevations
    D. renderings

7. Surface Hatches can be displayed in the following views: (select all that apply)

    A. model
    B. plan
    C. section
    D. elevation

8. Identify the tool shown.

    A. Expand Render Window
    B. Expand Tea Kettle
    C. Scale
    D. Render to Size

9. When there are no lights placed in a scene, the scene is rendered using:

    A. Default lighting.
    B. Sun & Sky.
    C. Nothing, it is really dark.
    D. Sun only, based on location.

10. Spotlights and point lights are represented by:

    A. Glyphs (symbols used to indicate the position and direction of the light)
    B. Nothing
    C. Blocks
    D. Linework

11. When mapping a material, you can adjust: (select all that apply)

    A. The X position
    B. The Y position
    C. Rotation
    D. Scale

12. Once you place a camera, you can adjust the camera using: (select all that apply)

  A. Properties
  B. The Adjust tool on the Render tab
  C. Using the MOVE and ROTATE tools
  D. You can't modify a camera once it has been placed

ANSWERS:

1) T; 2) T; 3) T; 4) T; 5) F; 6) A, B, C, D; 7) A, C, D; 8) D; 9) A; 10) A; 11) A, B, C, D; 12) A, B, C

Lesson 9:
Documentation

Before we can create our construction drawings, we need to create layouts or views to present the design. AutoCAD Architecture is similar to AutoCAD in that the user can work in Model and Paper Space. Model Space is where we create the 3D model of our house. Paper Space is where we create or set up various views that can be used in our construction drawings. In each view, we can control what we see by turning off layers, zooming, panning, etc.

To understand paper space and model space, imagine a cardboard box. Inside the cardboard box is Model Space. This is where your 3D model is located. On each side of the cardboard box, tear out a small rectangular window. The windows are your viewports. You can look through the viewports to see your model. To reach inside the window so you can move your model around or modify it, you double-click inside the viewport. If your hand is not reaching through any of the windows and you are just looking from the outside, then you are in Paper Space or Layout mode.

You can create an elevation in your current drawing by first drawing an elevation line and mark, and then creating a 2D or 3D elevation based on that line. You can control the size and shape of the elevation that is generated. Unless you explode the elevation that you create, the elevation remains linked to the building model that you used to create it. Because of this link between the elevation and the building model, any changes to the building model can be made in the elevation as well.

When you create a 2D elevation, the elevation is created with hidden and overlapping lines removed. You can edit the 2D elevation that you created by changing its display properties. The 2D Section/Elevation style allows you to add your own display components to the display representation of the elevation and create rules that assign different parts of the elevation to different display components. You can control the visibility, layer, color, linetype, lineweight, and linetype scale of each component. You can also use the line work editing commands to assign individual lines in your 2D elevation to display components and merge geometry into your 2D elevation.

After you create a 2D elevation, you can use the AutoCAD BHATCH and AutoCAD DIMLINEAR commands to hatch and dimension the 2D elevation.

AutoCAD Architecture comes with standard title block templates. They are located in the Templates subdirectory under *Program Data \ Autodesk\ ACA 2023 \enu \Template.*

Many companies will place their templates on the network so all users can access the same templates. In order to make this work properly, set the Options so that the templates folder is pointed to the correct path.

Exercise 9-1:

Creating a Custom Titleblock

Drawing Name: Architectural Title Block.dwg
Estimated Time: 30 minutes

This exercise reinforces the following skills:

- ❑ Title blocks
- ❑ Attributes
- ❑ Edit Block In-Place
- ❑ Insert Image
- ❑ Insert Hyperlink

1. 　📂　Select the **Open** tool.

2. | File name: | | |
   | Files of type: | Drawing Template (*.dwt) | |

   Set the Files of type to *Drawing Template*.

3. ProgramData
     Autodesk
       ACA 2020
         enu
           Template
             AutoCAD Templates

   Browse to
   ProgramData\Autodesk\ACA 2023\enu
   Template\AutoCAD Templates.

4. | File name: | Tutorial-iArch | |
   | Files of type: | Drawing Template (*.dwt) | |

   Open the *Tutorial-iArch.dwt.*
   Verify that *Files of type* is set to Drawing Template or
   you won't see the file.

5. Perform a **File→Save as**.

   | File name: | Arch_D|.dwt | |
   | Files of type: | AutoCAD Drawing Template (*.dwt) | |

   Save the file to your work folder.
   Rename **Arch_D.dwt**.

6.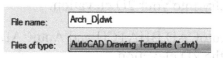

   Description
   Arch D

   Measurement
   English

   New Layer Notification
   ◉ Save all layers as unreconciled
   ◯ Save all layers as reconciled

   OK　Cancel　Help

   Enter a description and Click **OK**.

7. Firm Name and Address

    Zoom into the Firm Name and Address rectangle.

8. A
    Multiline
    Text

    Select the **MTEXT** tool from the **Annotation** panel on the Home ribbon.

    Draw a rectangle to place the text.

9. Set the Font to **Verdana**.

    Hint: If you type the first letter of the font, the drop-down list will jump to fonts starting with that letter.

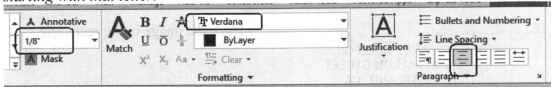

    Set the text height to ⅛".
    Set the justification to **centered**.
    Enter the name and address of your college.

10. LANEY COLLEGE
    800 FALLON STREET
    OAKLAND, CA

    Extend the ruler to change the width of your MTEXT box to fill the rectangle.

    Click **OK**.

11. Firm Name and Address

    LANEY COLLEGE
    800 FALLON STREET
    OAKLAND, CA

    Use **MOVE** to locate the text properly, if needed.

12. Home Insert Annotate Render V

    Attach Clip Adjust   Underlay Layers
                           Frames vary ▾
                           Snap to Underlays ON ▾
    Reference ▾

    Activate the **Insert** ribbon.

    Select the arrow located on the right bottom of the Reference panel.

13.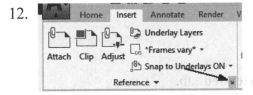

    Attach DWG...
    Attach Image...
    Attach DWF...

    The External Reference Manager will launch.

    Select **Attach Image** from the drop-down list.

14.  Locate the image file you wish to use.
There is an image file available for use from the publisher's website called *college logo.jpg*.
Click **Open**.

15. Click **OK**.

16. Place the image in the rectangle.

Firm Name and Address

LANEY COLLEGE
800 FALLON STREET
OAKLAND, CA

Laney College

Tips & Tricks

If you are concerned about losing the link to the image file (for example, if you plan to email this file to another person), you can use **INSERTOBJ** to embed the image into the drawing. Do not enable link to create an embedded object. The INSERTOBJ command is not available on the standard ribbon.

---

17. Select the image.
Right click and select **Properties**.

18. Select the **Extended** tab.
Pick the Hyperlink field.

Raster Image	
DOCUMENTATION	
Hyperlink	
Notes	
Reference documents	(0)

19.

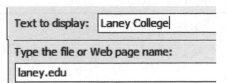

Type in the name of the school in the Text to display field.

Type in the website address in the *Type the file or Web page name* field.

Click **OK**.
Close the Properties dialog.

20.

If you hover your mouse over the image, you will see a small icon. This indicates there is a hyperlink attached to the image. To launch the webpage, simply hold down the CTRL key and left click on the icon.

If you are unsure of the web address of the website you wish to link, use the Browse for **Web Page** button.

21.

To turn off the image frame/boundary, type **IMAGEFRAME** at the command line. Enter **0**.

IMAGEFRAME has the following options:

        0: Turns off the image frame and does not plot

        1: Turns on the image frame and is plotted

        2: Turns on the image frame but does not plot

22.

Block Editor
Edit Block in-place...
Edit Attributes...

Select the titleblock.
Right click and select **Edit Block in-place**.

23. 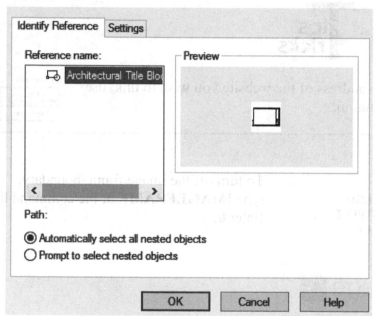 Click **OK**.

24.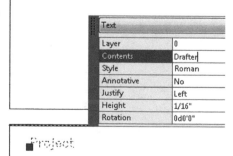

Change the text for Project to **Drafter**.
To change, simply double click on the text and an edit box will appear.

25. 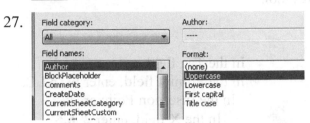 Select the **Define Attribute** tool from the Insert ribbon.

26. Select the **Field** tool.

27. 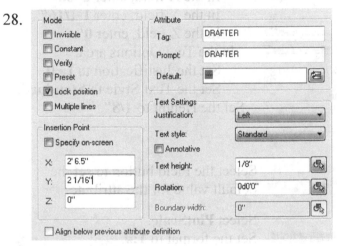 Highlight Author and Uppercase.

Click **OK**.

28. In the Tag field, enter **DRAFTER**.
In the Prompt field, enter **DRAFTER**.
The Value field is used by the FIELD property.
  In the Insertion Point area:
    In the X field, enter **2′ 6.5″**.
    In the Y field, enter **2-1/16″**.
    In the Z field, enter **0″**.
  In the Text Options area:
    Set the Justification to **Left**.
    Set the Text Style to **Standard**.
    Set the Height to **1/8″**.

Click **OK**.

29. 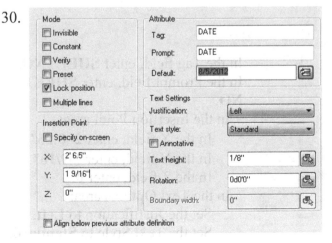 Select the **Define Attributes** tool.

30. In the Tag field, enter **DATE**.
In the Prompt field, enter **DATE**.

  In the Insertion Point area:
    In the X field, enter **2′ 6.5″**.
    In the Y field, enter **1-9/16″**.
    In the Z field, enter **0″**.
  In the Text Options area:
    Set the Justification to **Left**.
    Set the Text Style to **Standard**.
    Set the Height to **1/8″**.

31. Select the Field button to set the default value for the attribute.

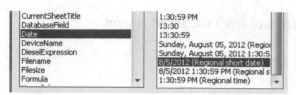

Select **Date**.
Set the Date format to **Regional short date**.

Click **OK**.

32.  Select the **Define Attributes** tool.

33.

In the Tag field, enter **SCALE**.
In the Prompt field, enter **SCALE**.
  In the Insertion Point area:
    In the X field, enter **2′ 6.5″**.
    In the Y field, enter **1-1/16″**.
    In the Z field, enter **0″**.
  In the Text Options area:
    Set the Justification to **Left**.
    Set the Text Style to **Standard**.
Set the Height to **1/8″**.

34.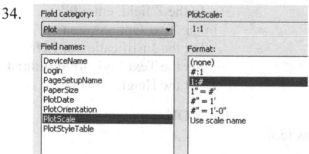

Select the Field button to set the default value for the attribute.

Select **PlotScale**.
Set the format to **1:#**.

Click **OK**.

Click **OK** to place the attribute.

35. 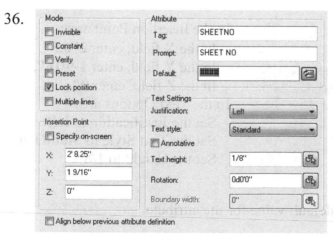 Select the **Define Attributes** tool.

36.

In the Tag field, enter **SHEETNO**.
In the Prompt field, enter **SHEET NO**.
  In the Insertion Point area:
    In the X field, enter **2′ 8.25″**.
    In the Y field, enter **1-9/16″**.
    In the Z field, enter **0**.
  In the Text Options area:
    Set the Justification to **Left**.
    Set the Text Style to **Standard**.
    Set the Height to **1/8″**.

37. Select the Field button to set the default value for the attribute.

Select **CurrentSheetNumber**. Set the format to **Uppercase**.

Click **OK**.

Click **OK** to place the attribute.

38. Select the **Define Attribute** tool.

Define
Attributes

39.  In the Tag field, enter **PROJECTNAME**.
In the Prompt field, enter **PROJECT NAME**.
In the Insertion Point area:
In the X field, enter **2′ 7-9/16″**.
In the Y field, enter **4″**.
In the Z field, enter **0**.
In the Text Options area:
Set the Justification to **Center**.
Set the Text Style to **Standard**.
Set the Height to **1/8″**.

40. Select the Field button to set the default value for the attribute.

 Select **AEC Project** for the field category.
Set the Field Name to **Project Name**.
Set the format to **Uppercase**.

Click **OK**.

A common error for students is to forget to enter the insertion point. The default insertion point is set to 0,0,0. If you don't see your attributes, look for them at the lower left corner of your title block and use the MOVE tool to position them appropriately.

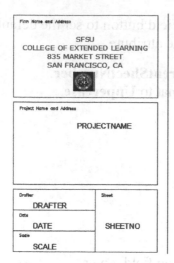

Your title block should look similar to the image shown.

If you like, you can use the MOVE tool to reposition any of the attributes to make them fit better.

41. Select **Save** on the ribbon to save the changes to the title block and exit the Edit Block In-Place mode.

42. Click **OK**.

AutoCAD ✕

⚠ All references edits will be saved.

- To save reference changes, click OK.
- To cancel the command, click Cancel.

[OK]   [Cancel]

43. Save the file and go to **File→Close**.

Tips & Tricks

➢ Use templates to standardize how you want your drawing sheets to look. Store your templates on a server so everyone in your department uses the same titleblock and sheet settings. You can also set up dimension styles and layer standards in your templates.

➢ ADT 2005 introduced a new tool available when you are in Paper Space that allows you to quickly switch to the model space of a viewport without messing up your scale. Simply select the **Maximize Viewport** button located on your task bar to switch to model space.

➢ To switch back to paper space, select the **Minimize Viewport** button.

You use the Project Navigator to create additional drawing files with the desired views for your model. The views are created on the Views tab of the Project Navigator. You then create a sheet set which gathers together all the necessary drawing files that are pertinent to your project.

Previously, you would use external references, which would be external drawing files that would be linked to a master drawing. You would then create several layout sheets in your master drawing that would show the various views. Some users placed all their data in a single drawing and then used layers to organize the data.

This shift in the way of organizing your drawings will mean that you need to have a better understanding of how to manage all the drawings. It also means you can leverage the drawings so you can reuse the same drawing in more than one sheet set.

You can create five different types of views using the Project Navigator:

- Model Space View – a portion that is displayed in its own viewport. This view can have a distinct name, display configuration, description, layer snapshot and drawing scale.
- Detail View – displays a small section of the model, i.e. a wall section, plumbing, or foundation. This type of view is usually associated with a callout. It can be placed in your current active drawing or in a new drawing.
- Section View – displays a building section, usually an interior view. This type of view is usually associated with a callout. It can be placed in your current active drawing or in a new drawing.
- Elevation View – displays a building elevation, usually an exterior view. This type of view is usually associated with a callout. It can be placed in your current active drawing or in a new drawing.
- Sheet View – this type of view is created when a model space view is dropped onto a layout sheet.

Note: Some classes have difficulty using the Project Navigator because they do not use the same workstation each class or the drawings are stored on a network server. In those cases, the links can be lost and the students get frustrated trying to get the correct results.

As we start building our layout sheets and views, we will be adding schedule tags to identify different building elements.

Schedule tags can be project-based or standard tags. Schedule tags are linked to a property in a property set. When you anchor or tag a building element, the value of the property displays in the tag. The value displayed depends on which property is defined by the attribute used by the schedule tag. ACA comes with several pre-defined schedule tags, but you will probably want to create your own custom tags to display the desired property data.

Property sets are specified using property set definitions. A property set definition is a documentation object that is tracked with an object or object style. Each property has a name, description, data type, data format, and default value.

To access the property sets, open the Style Manager. Highlight the building element you wish to define and then select the Property Sets button. You can add custom property sets in addition to using the default property sets included in each element style.

To determine what properties are available for your schedules, select the element to be included in the schedule, right click and select Properties.

Tags and schedules use project information. If you do not have a current project defined, you may see some error messages when you tag elements.

Elevation Views

Elevations of the building models are created by using the Elevation tool on the Annotate ribbon. You first draw an elevation line and mark to indicate the direction of the elevation, selecting the scope/depth of the view, and then placing a 2D or 3D elevation based on the elevation line. You can control the size and shape of any elevation that you create. If you modify the building model, you can update the elevation view so it shows the changes. 2D elevations are created with hidden and overlapping lines removed. You can control the display of 2D elevations by modifying or creating elevation styles and modifying the display properties of the 2D elevation.

Exercise 9-2:

Creating Elevation Views

Drawing Name: elevation.dwg
Estimated Time: 15 minutes

This exercise reinforces the following skills:

    ❑ Creating an Elevation View
    ❑ Named Views

1.   Open *elevation.dwg*.

2.   Switch to the Model tab.

3.   Activate the **Top** view by clicking the top plane on the view cube. *You can also use the shortcut menu in the upper left corner of the display window to switch views.*

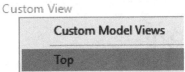

4.   Verify using the View cube that North is oriented on top.

5.   [–][Top][2D Wireframe]   Set the view to **2D Wireframe** mode by selecting the View Display in the upper left corner of the window.

6.   Zoom Window / Zoom Extents / Zoom Previous

Use **Zoom Extents** to view the entire model.

7.   Activate the **Annotate** ribbon.

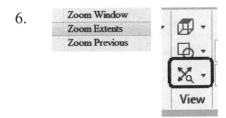

Select the **Elevation** tool on the Callouts panel.

8. Place the elevation mark below the model.
Use your cursor to orient the arrow toward the building model.

9. Set your view name to **South Elevation**.

Enable **Generate Section/Elevation**.

Enable **Place Titlemark**.

Set the Scale to **1/8″ = 1′-0″**.

Click the **Current Drawing** button.

Callout Only

New Model Space View Name:

South Elevation

Create in:

New View Drawing

Existing View Drawing

Current Drawing

☑ Generate Section/Elevation
☑ Place Titlemark
Scale: 1/8" = 1'-0"

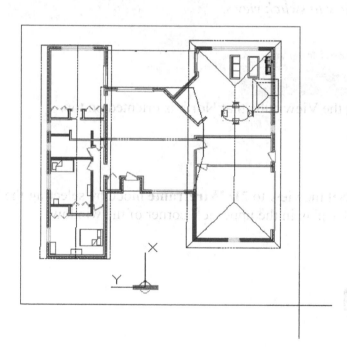

10. Window around the entire building to select it.
Select the upper left corner above the building and the lower right corner below the building.

11. 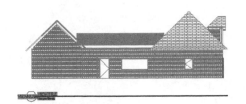 Place the elevation to the right of the view.

Zoom into the elevation view, so you can inspect it.

12.  Switch to the View tab.

Note that the elevation view you just created is listed.

13. Save as *ex9-2.dwg*.

Layouts

Layouts act as the sheets for your documentation set. You name the layouts, add views to the layouts, add text and dimensions, and any other relevant annotation.

You place one or more viewports on a layout. The area outside the viewport is called *paper space*. Normally, paper space is used to add a title block and possibly notes. Views are created in model space for any schedules, elevations, plans, etc. These are then added to the layout. The viewport controls the scale and display of the model elements inside the viewport.

You can access one or more layouts from the tabs located at the bottom-left corner of the drawing area to the right of the Model tab. You can use multiple layout tabs to display details of the various components of your model at several scales and on different sheet sizes.

Exercise 9-3:
Creating Layouts

Drawing Name: layouts.dwg
Estimated Time: 25 minutes

This exercise reinforces the following skills:

- Create Layout
- Adding a View to a Layout
- Control Visibility of elements using Layers

1.   Open *layouts.dwg.*

2.   Right click on the command line and select **Options**.

    You can also type OPTIONS.

3.   Activate the **Display** tab.

    Enable **Display Layout and Model tabs**.

    Click **OK**.

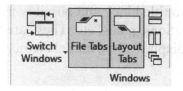

These can also be enabled on the View ribbon.

4.   Right click on the Work tab.

    Select **From template**.

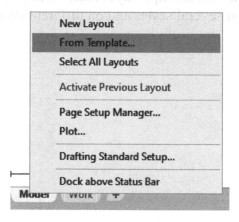

5. Browse to your exercise folder.

    Select the *Arch_D* template.
    Click **Open**.

    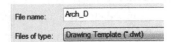

    | File name: | Arch_D |
    | Files of type: | Drawing Template (*.dwt) |

6. | Layout name(s): | OK |
    | D-Size Layout | Cancel |

    Click **OK**.

7. Model / Work \ D-Size Layout /   Activate the **D-Size Layout**.

8. From Template...
    Delete
    Rename
    Move or Copy...
    Select All Layouts

    Activate Previous Layout
    Activate Model Tab

    Page Setup Manager...
    Plot...

    Drafting Standard Setup...

    Import Layout as Sheet...
    Export Layout to Model...

    Dock above Status Bar

    Size Layout

    Right click on the **D-Size Layout** tab.

    Select **Rename**.

9. \ Floorplan /   Rename **Floorplan**.
    Activate/Open the **Floorplan** layout.

10. | Misc | |
    | On | Yes |
    | Clipped | No |
    | Display locked | No |
    | Annotation scale | 1'-0" = 1'-0" |
    | Standard scale | 1/4" = 1'-0" |

    Select the viewport.
    *If the Properties dialog doesn't come up, right click and select **Properties**.*

    Set Display locked to **No**.

11.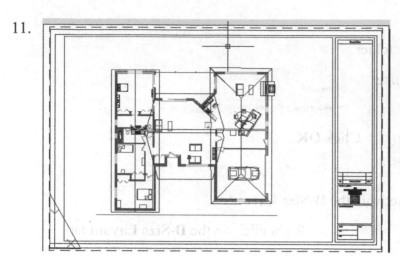

Resize the viewport to only show the floor plan.

Double click inside the viewport to activate model space.

Pan the model to position inside the viewport.

12.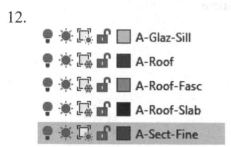

Activate the Home ribbon.

Freeze the layers assigned to roofs and roof slabs in the current viewport.

13.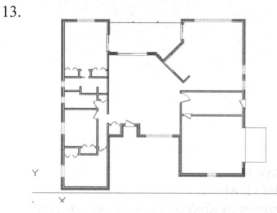

The view updates with no roof.

14.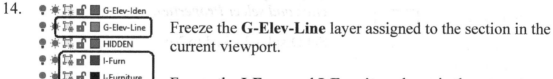

Freeze the **G-Elev-Line** layer assigned to the section in the current viewport.

Freeze the **I-Furn** and **I-Furniture** layer in the current viewport.

15. If you see the camera:

To turn off the display of the camera, type **CAMERADISPLAY**, and then **0.**

16.

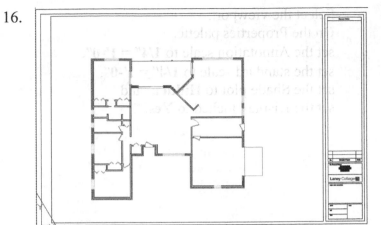

Double click outside the viewport to return to paper space.

Observe how the floor plan view has changed.

17.

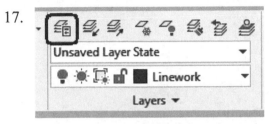

Launch the Layer Manager.

18.

Scroll down to the Viewport layer. Set the Viewport layer to DO NOT PLOT.

The viewports can be visible so you can adjust them but will not appear in prints.

19.

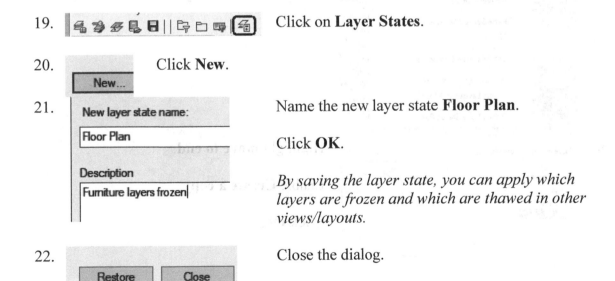

Click on **Layer States**.

20.

New...

Click **New**.

21.

New layer state name:

Floor Plan

Description

Furniture layers frozen|

Name the new layer state **Floor Plan**.

Click **OK**.

By saving the layer state, you can apply which layers are frozen and which are thawed in other views/layouts.

22.

Restore   Close

Close the dialog.

Close the Layer Manager dialog.

23.

Layer	Viewport
Display locked	Yes
Annotation scale	1/4" = 1'-0"
Standard scale	1/4" = 1'-0"
Custom scale	1/32"
Layer property o...	No
Visual style	2D Wireframe
Shade plot	Hidden

Select the viewport.
On the Properties palette,
set the Annotation scale to **1/4″ = 1′-0″**,
set the standard scale to **1/4″ = 1′-0″**,
set the Shade plot to **Hidden**, and
set the Display locked to **Yes**.

24. Save as *ex9-3.dwg*.

Exercise 9-4:
Copy a Layout

Drawing Name: copy_layout.dwg
Estimated Time: 15 minutes

This exercise reinforces the following skills:

- ❑ Layouts
- ❑ Viewports

1. Open *copy_layout.dwg*.

2. 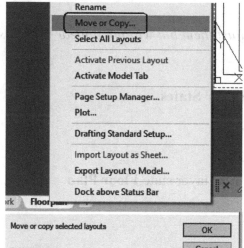 Right click on the Floorplan layout.

   Select **Move or Copy**.

3. Highlight **move to end**.

   Enable **Create a copy**.

   Click **OK**.

4.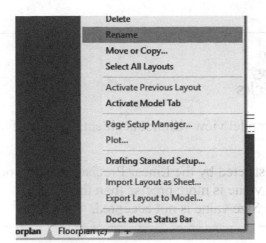

Right click on the copied layout.

Select **Rename**.

5. 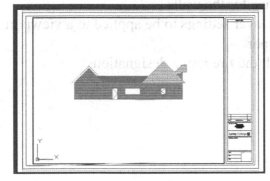 Rename **South Elevation**.

6. Activate/Open the South Elevation layout.
Select the viewport.
Right click and set the Display locked to **No**.

7. Double click inside the viewport to activate model space.

8.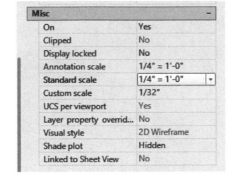

Left click inside the viewport and then position the elevation view inside the viewport.

Use PAN to move the view over in the model.

9.

Misc	–
On	Yes
Clipped	No
Display locked	No
Annotation scale	1/4" = 1'-0"
Standard scale	1/4" = 1'-0"
Custom scale	1/32"
UCS per viewport	Yes
Layer property overrid...	No
Visual style	2D Wireframe
Shade plot	Hidden
Linked to Sheet View	No

Select the viewport.
Set the Annotation scale to **1/4" = 1'-0" [1:100]**.
Set the Standard scale to **1/4"= 1'-0" [1:100]**.

Set the Shade plot to **Hidden**.

You may need to re-center the view in the viewport after you change the scale. Use PAN to do this so the scale does not change.

Lock the display.

10. Save the file as *ex9-4.dwg*.

> ➢ By locking the Display you ensure your model view will not accidentally shift if you activate the viewport.

> ➢ The Annotation Plot Size value can be restricted by the Linear Precision setting on the Units tab. If the Annotation Plot Size value is more precise than the Linear Precision value, then the Annotation Plot Size value is not accepted.

Most projects require a floorplan indicating the fire ratings of the building walls. These are submitted to the planning department for approval to ensure the project is to building code. In the next exercises, we will complete the following tasks:

- Load the custom linetypes to be used for the exterior and interior walls
- Apply the 2-hr linetype to the exterior walls using the Fire Rating Line tool
- Apply the 1-hr linetype to the interior walls using the Fire Rating Line tool
- Add wall tags which display fire ratings for the walls
- Create a layer group filter to save the layer settings to be applied to a viewport
- Apply the layer group filter to a viewport
- Create the layout of the floorplan with the fire rating designations

Exercise 9-5:
Loading a Linetype

Drawing Name: linetypes.dwg
Estimated Time: 10 minutes

This exercise reinforces the following skills:

- Loading Linetypes

FIRE RATED, SMOKE BARRIER WALLS, 1 HOUR

FIRE RATED, SMOKE BARRIER WALLS, 2 HOUR

FIRE RATED, SMOKE BARRIER WALLS, 3 HOUR

FIRE RATED, SMOKE BARRIER WALLS, 4 HOUR

These are standard line patterns used to designate different fire ratings.

I have created the linetypes and saved them into the firerating.lin file included in the exercise files. You may recall we used the ExClick Tools Make Linetype tool to create custom linetypes.

1. Open *linetypes.dwg*.

2.  Activate the Model tab.

3. Type **LINETYPE** to load the linetypes.
4. 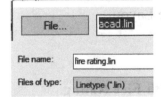 Select **Load**.
5.
    Select the **File** button to load the linetype file.
6. Locate the *fire rating.lin* file in the exercise files.

    Click **Open**.

7.  *This is a txt file you can open and edit with Notepad.* Hold down the control key to select both linetypes available.

    Click **OK** to load.
8. 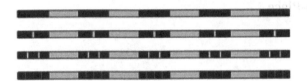 The linetypes are loaded.

    Click **OK** to close the dialog box.
9. Save as *ex9-5.dwg*.

Exercise 9-6:
Applying a Fire Rating Line

Drawing Name: fire rating.dwg
Estimated Time: 25 minutes

This exercise reinforces the following skills:

- ❑ Adding a Layout
- ❑ Layers
- ❑ Edit an External Reference In-Place
- ❑ Adding Fire Rating Lines

1.   Open *fire rating.dwg*.

2.   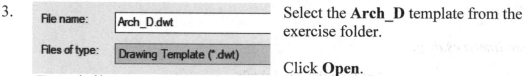 Right click on the + layout tab.

    Select **From Template**.

3.   Select the **Arch_D** template from the exercise folder.

    Click **Open**.

4.   Highlight the Layout name.

    Click **OK**.

5.   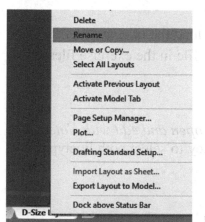 Right click on the D-Size Layout.

    Select **Rename**.

    Rename to **Floorplan with Fire Rating**.

6.   Open/activate the **Floorplan with Fire Rating** layout.

7.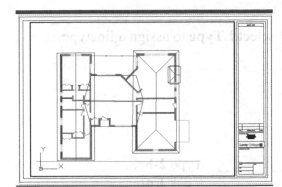

Double click inside the layout to activate model space.

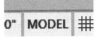

Note the viewport is bold when model space is activated. You also see MODEL on the status bar.

8.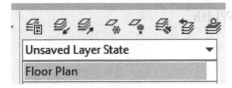

Set the Layer State to **Floor Plan**.

This will freeze the roof and furniture layers.

9.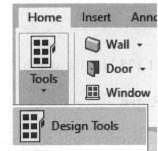

Launch the **Design Tools** palette from the Home ribbon if it is not available.

10.

Right click on the Design Tools palette title bar.

Enable **Document**.

11.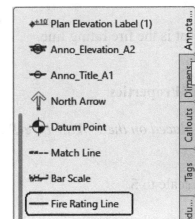

Select the Annotation tab.

Locate and select the **Fire Rating Line** tool.

12.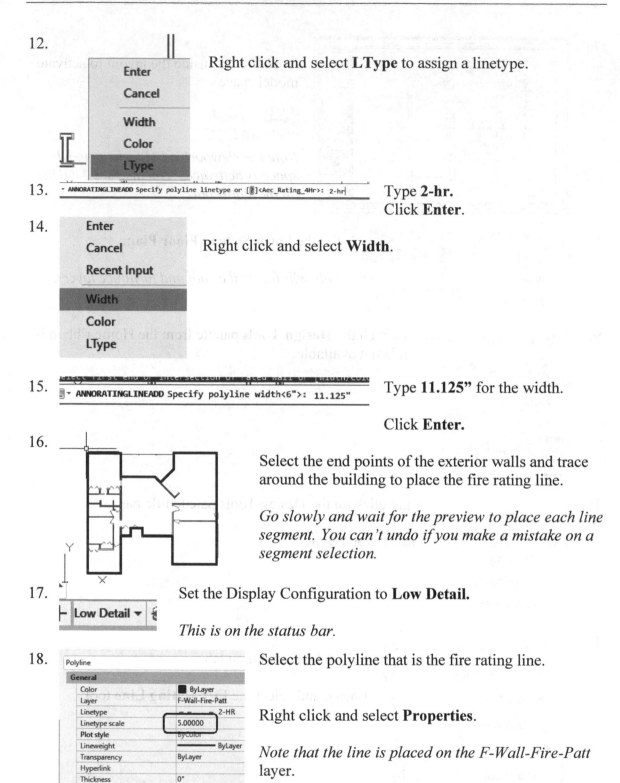

Right click and select **LType** to assign a linetype.

13. `ANNORATINGLINEADD Specify polyline linetype or [?]<Aec_Rating_4Hr>: 2-hr`

Type **2-hr.**
Click **Enter.**

14.

Right click and select **Width.**

15. `ANNORATINGLINEADD Specify polyline width<6">: 11.125"`

Type **11.125"** for the width.

Click **Enter.**

16.

Select the end points of the exterior walls and trace around the building to place the fire rating line.

Go slowly and wait for the preview to place each line segment. You can't undo if you make a mistake on a segment selection.

17. Low Detail ▾

Set the Display Configuration to **Low Detail.**

This is on the status bar.

18.

Select the polyline that is the fire rating line.

Right click and select **Properties.**

Note that the line is placed on the F-Wall-Fire-Patt layer.

Change the linetype scale to **5.**

Click **ESC** to release the selection.

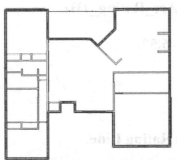

You should be able to see the hatch pattern of the fire rating line.

19.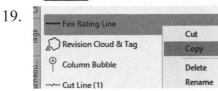

Select the **Fire Rating Line** tool.
Right click and select **Copy**.

20.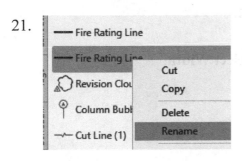

Right click on the palette.

Click **Paste**.

21.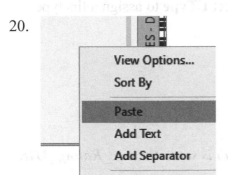

Select the duplicate Fire Rating Line tool.

Right click and select **Rename**.

22.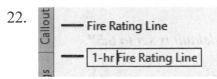

Rename **1-hr Fire Rating Line**.

23.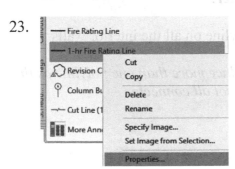

Highlight the **1-hr Fire Rating Line**.

Right click and select **Properties.**

24.

BASIC	
General	
Layer key	⊞ WALLFIRE
Layer overrid...	⊞ --
Line width	6 1/2"
Linetype	Aec_Rating_1Hr

Set the Linetype to **Aec_Rating_1Hr.**

Set the Linewidth to **6.5".**

Click **OK.**

25.

— Fire Rating Line

— 1-hr Fire Rating Line

☁ Revision Cloud & Tag

Select the **1-hr Fire Rating Line.**

26.

Enter
Cancel
Width
Color
LType

Right click and select **LType** to assign a linetype.

27.

or [**?**]<Aec_Rating_1Hr>: |

*Notice the default is set to the **Aec_Rating_1Hr.***

Click **ENTER.**

28.

Enter
Cancel
Recent Input
Width
Color
LType

Right click and select **Width.**

29.

Specify polyline width<6 1/2">:

*Notice the default is set to **6.5".***

Click **Enter.**

30.

Place the 1-hr polyline on all the interior walls.

You will need to place more than one polyline as the interior walls are not all connected.

31.

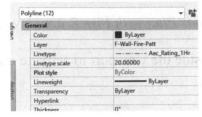

Type **qslelect.**

Select **Polyline** from the Object Type drop-down list.

Highlight **Linetype** in Properties.

Set the Value to **Aec_Rating_1Hr.**

Click **OK.**

32.

Right click and select **Properties**.

Set the Linetype scale to **20.00.**

Click ESC to release the selection.

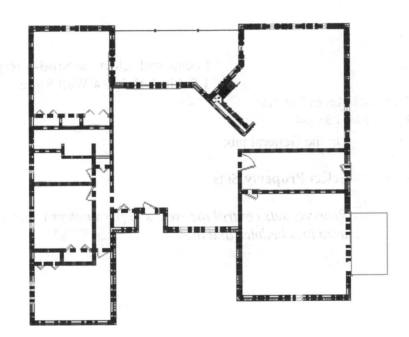

33.   Click outside the viewport to return to paper space.

34.   Save as *ex9-6.dwg*.

Exercise 9-7:
Assigning Fire Ratings to Wall Properties

Drawing Name: fire rating walls.dwg
Estimated Time: 10 minutes

This exercise reinforces the following skills:

    ❑   Wall Properties

1.   Open *fire rating walls.dwg*.

2.   Activate the *Floor Plan with Fire Rating* layout.

    Double click inside the viewport to activate model space.

3.   Activate the Manage ribbon.

    Select the **Style Manager**.

4.   Locate and select the **Stud-4 Rigid-1.5 Air-1-Brick-4** Wall Style.

    Wall Styles
    Brick_Block
    Standard
    Stud-4 GWB-0.625 2 Layers Each Side
    Stud-4 Rigid-1.5 Air-1 Brick-4

5.   On the General tab:

    Select **Property Sets**.

    Property sets control the values of parameters which are used in schedules and tags.

General | Components | Materials | Endcaps / Opening Endcaps | Classifi

Name:
Stud-4 Rigid-1.5 Air-1 Brick-4

Description:
4" Stud Cavity Wall (Actual) with 4" Brick and 1.5" Rigid

☑ Objects of this style may act as a boundary for associative spaces

Notes... | Property Sets...

6. 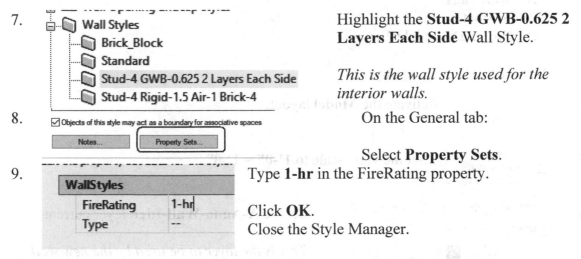   Type **2-hr** in the FireRating property.

Click **OK**.

When defining the attribute to use this value, we use the format WallStyles:FireRating. The first word is the property set and the second word is the property.

7.   Highlight the **Stud-4 GWB-0.625 2 Layers Each Side** Wall Style.

This is the wall style used for the interior walls.

8.   On the General tab:

Select **Property Sets**.

9.   Type **1-hr** in the FireRating property.

Click **OK**.
Close the Style Manager.

10. Double click outside the viewport to return to paper space.

Save the file as *ex9-7.dwg*.

Property Sets

Schedules use data from property sets. By adding the Fire Rating information to the wall styles, you can use that data in any schedules. You can create custom properties for your styles to leverage your ACA elements in schedules.

Locks may appear on some property set definitions and property definitions when you open legacy engineering drawings. Data that is programmatically set cannot be modified and is protected and identified by locks:

- selections on the Applies To tab and names for property set definitions

- anything that affects the value of underlying data and names for property definitions

Exercise 9-8:
Creating a Schedule Tag

Drawing Name: wall schedule_tag.dwg
Estimated Time: 15 minutes

This exercise reinforces the following skills:

- ❑ Wall Tags
- ❑ Attributes
- ❑ Create a Schedule Tag

1.   Open *wall schedule tag.dwg*.

2.   Activate the **Model** layout.

3.   Set the view scale to **1'-0" = 1'-0"**.

4.   Set the **A-Anno-Wall-Iden** layer current.

        This is the layer to be used by the new wall tag.

5.   Draw an 8" x 8" square.

        Hint: *Use the Rectangle tool. Click one corner and for the second corner, type @ 8", 8".*

6.   Select the square.
        Right click and select **Basic Modify Tools→Rotate**.

        When prompted for the base point, right click and select **Geometric Center**.

        Then select the center of the square.

        Type **45** for the angle of rotation.

7. Activate the Insert ribbon.

Select the **Define Attributes** tool.

8.

Attribute	
Tag:	WALLSTYLES:FIRERATING
Prompt:	Fire Rating
Default:	2-hr

Text Settings	
Justification:	Middle center
Text style:	RomanS
☐ Annotative	
Text height:	1"
Rotation:	0.00
Boundary width:	0"

Type
WALLSTYLES:FIRERATING for the tag.
This is the property set definition used for this tag.
Type **Fire Rating** for the prompt.
Type **2-hr** for the Default value.

Set the Justification for the **Middle cente**r.
Set the Text style to **RomanS**.
Disable **Annotative**.
Set the Text Height to **1"**.

Enable **Specify on-screen**.

Click **OK**.

9. Select the Geometric Center of the Square to insert the attribute.

10. *In order to use annotative scales, we need to re-size these elements.*

Use the modify tools to scale the square and the attribute by a factor of 10.67.

11.

Door Tag ▾	Door Schedule ▾
Window Tag	Window Schedule
Room Tag ▾	Room Schedule ▾

Wall Tag (Leader)	Wall Schedule
Create Tag	Schedule Styles
Renumber Property Sets	Evaluate Space
Scheduling	

Activate the Annotate ribbon.

Select the **Create Tag** tool from the Scheduling panel in the drop-down area.

12.

Select the square and the attribute you just created.

Press **ENTER**.

13.

Name the new tag **Wall Fire Rating - Imperial**.

You should see a preview of the tag and you should see the Property Set and Property Definition used by the tag based on the attribute you created.

Click **OK**.

14.

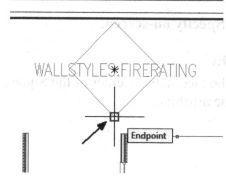

You will be prompted for the insertion point for the tag.

Select the bottom vertex of the rectangle.

15.

The tag will convert to a block and you will see the default attribute.

Notice that when it converts, the size will change.

If you don't see the tag, check that the Display Property is set to Medium Detail on the status bar.

16. Save the drawing as *ex9-8.dwg*.

Exercise 9-9:
Creating a Schedule Tag Tool on the Document Palette

Drawing Name: palette tools.dwg
Estimated Time: 10 minutes

This exercise reinforces the following skills:

- Wall Tags
- Create a Tool on the Tool Palette
- Property Sets

1. Open *palette tools.dwg*.

2. 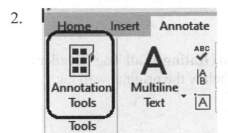 Switch to the Annotate ribbon.

             Launch the **Annotation Tools**.

3. 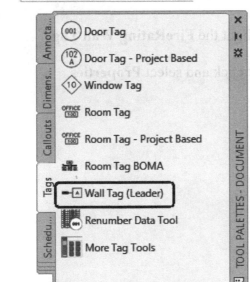 Click on the **Tags** tab of the Tool Palette.

             Locate the **Wall Tag (Leader) tool** on the Document tool palette.

4.   Highlight the **Wall Tag (Leader) tool**.
Right click and select **Copy**.

5. 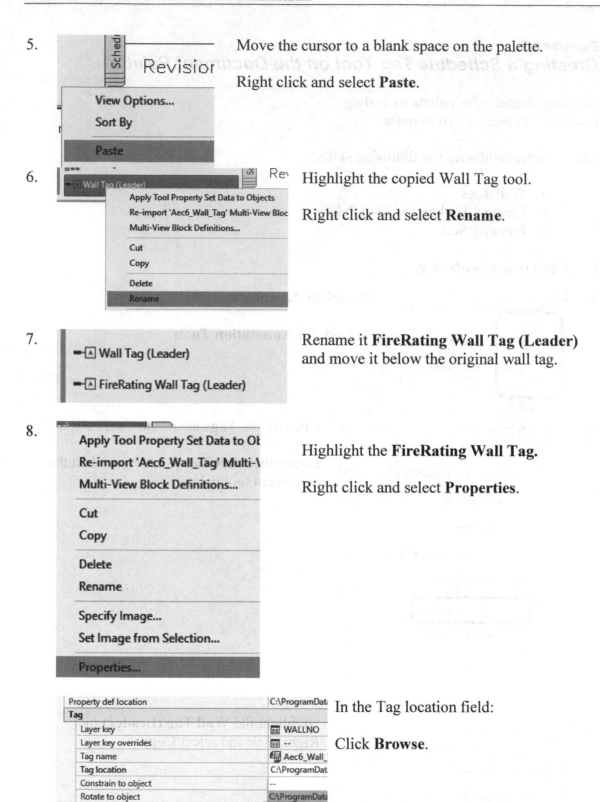 Move the cursor to a blank space on the palette.

Right click and select **Paste**.

6. Highlight the copied Wall Tag tool.

Right click and select **Rename**.

7. Rename it **FireRating Wall Tag (Leader)** and move it below the original wall tag.

8. Highlight the **FireRating Wall Tag.**

Right click and select **Properties**.

In the Tag location field:

Click **Browse**.

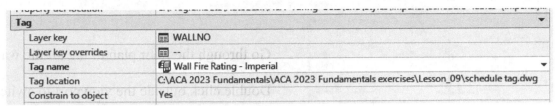

Tag		
Layer key	WALLNO	
Layer key overrides	--	
Tag name	Wall Fire Rating - Imperial	
Tag location	C:\ACA 2023 Fundamentals\ACA 2023 Fundamentals exercises\Lesson_09\schedule tag.dwg	
Constrain to object	Yes	

9. Set the Tag Location as *schedule_tag.dwg*.
   Set the Tag Name to **Wall Fire Rating - Imperial**.

   You may want to create a drawing to store all your custom schedule tags and then use that drawing when setting the tag location. In this case, I created the drawing and loaded the tag we created in the previous exercise.

   Click **OK**.

10. 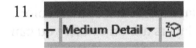   Select the **Floorplan with Fire Rating** layout.

    Activate **Model** space by clicking inside the viewport.

11. ── Medium Detail ▾   Set the Display to **Medium Detail**.

12.   FireRating Wall Tag (Leader)   Select the **FireRating Wall Tag** tool and tag an exterior wall.

    Click **ENTER** to complete the placement.
    Click **OK** to accept the data set.

13. Edit the property set data for the object:

PROPERTY SETS FROM STYLE	▾
WallStyles	▾
Data source	C:\ACA 2023 Fund...
FireRating	2-HR
Type	--

14.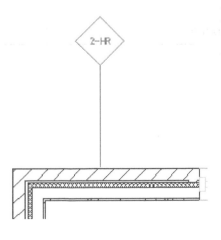

    Left click to place the tag and then click **OK** to accept the properties.

    You should see the updated tag.

15.

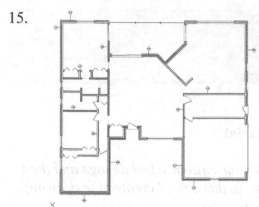

Go through the floor plan to tag all the walls.

Double click outside the viewport to switch to paper space.

16. Save as *ex9-9.dwg*.

Layer States

Layer states are used to save configurations of layer properties and their state (ON/OFF/FROZEN/LOCKED). For example, you can set a layer to be a different color depending on the state. Instead of constantly resetting which layers are on or off, you can save your settings in layer states and apply them to layouts.

Layer States are saved in the drawing. To share them across drawings, you need to export them. Each layer state has its own LAS file. To export a layer state, select it in the Layer States Manager and choose Export.

To import a layer state, open the Layer States Manager and click the Import tool. Then select the *.las file you wish to import.

You can save layer states in your drawing template.

Exercise 9-10:
Modify a Dimension Style

Drawing Name: dim_style.dwg
Estimated Time: 15 minutes

This exercise reinforces the following skills:

    ❑ Dimension Styles
    ❑ AEC Dimensions

1.   Open *dim_style.dwg*.

2.   Activate the Annotate ribbon.

    Select the **Dimension Style Editor** on the Dimensions panel.

3.   Highlight **Annotative**.

4.   Select **Modify**.

5.   On the Lines tab,
    set the Baseline spacing to **6"**.

6.   Set Extend beyond dim lines to **2"**.
    Set Offset from origin to **1"**.

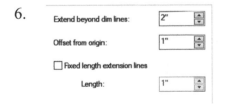

7.   On the Symbols and Arrows tab,

    Select **Architectural** tick for the arrowheads.
    Set the Arrow size to **6"**.

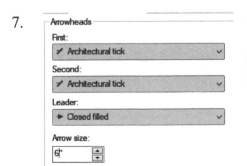

8. 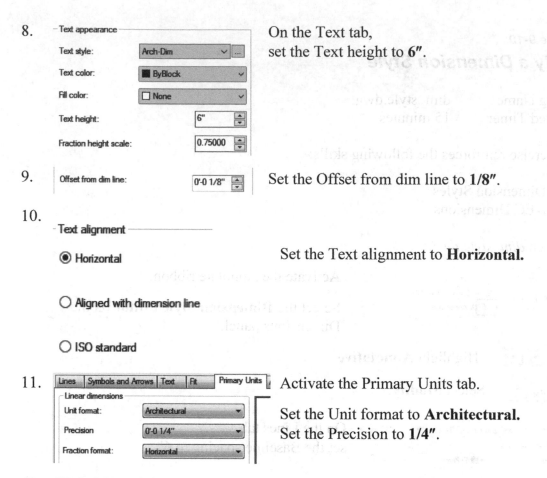 On the Text tab,
set the Text height to **6″**.

9. Set the Offset from dim line to **1/8″**.

10.

Set the Text alignment to **Horizontal.**

11. Activate the Primary Units tab.

Set the Unit format to **Architectural.**
Set the Precision to **1/4″**.

12. Click **OK** and **Close**.

13. Save as *ex9-10.dwg*.

Exercise 9-11:

Dimensioning a Floor Plan

Drawing Name: dim1.dwg
Estimated Time: 30 minutes

This exercise reinforces the following skills:

- ❑ Drawing Setup
- ❑ AEC Dimensions
- ❑ Title Mark
- ❑ Attributes
- ❑ Layouts

1.   Open *dim1.dwg*.

2.   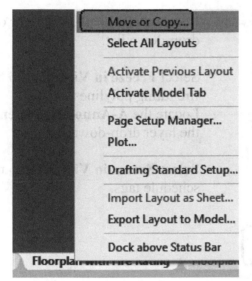   Highlight the **Floor Plan with FireRating** layout tab.

Right click and select **Move or Copy**.

3.      Enable **Create a copy**.

Highlight **(move to end)**.

Click **OK**.

4.   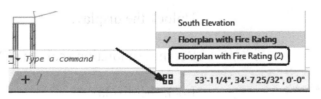   *If you don't see the copied layout:*

Left click on the layouts icon on the Status Bar.

5. 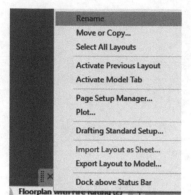 Highlight the copied layout tab.

Right click and select **Rename**.

6. 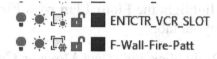 Modify the layout name to **A102 02 FLOOR PLAN**.

7. MODEL   Activate model space for the new layout.

Activate the Home ribbon.

8. ENTCTR_VCR_SLOT

F-Wall-Fire-Patt

Locate the **F-Wall-Patt** layer on the layer drop-down list.

Select **Freeze in Viewport** to turn off the fire rating polylines.

9. 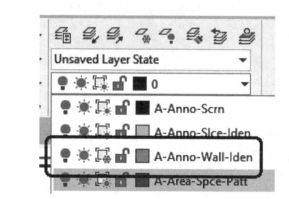 Locate the **A-Anno-Wall-Iden** layer on the layer drop-down list.

Select **Freeze in Viewport** to turn off the schedule tags.

10.

Viewport	
Layer	Viewport
Display locked	No
Annotation scale	1/4" = 1'-0"
Standard scale	1/4" = 1'-0"
Custom scale	1/32"
Layer property o...	No
Visual style	2D Wireframe
Shade plot	Wireframe

Return to PAPER SPACE.
(Click outside the viewport.)

Select the viewport.

Unlock the display.

Set the Annotation scale to ¼" = 1'-0".

Lock the display.

Click **ESC** to release the selected viewport.

11.  Note that you are in **PAPER** space on the layout.

12. 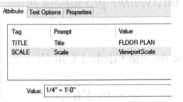 Activate the Annotate ribbon.

Select the **Title Mark** tool.
Pick below the view to place the title mark.

13. VIEWTITLE

ViewportScale

Pick two points to indicate the start and end points of the title mark.

Place the title mark below the view.

14. 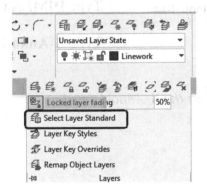 Double click to edit the attributes.
In the Title field, enter **FLOOR PLAN**.
In the SCALE field, enter **¼″ = 1′-0″**.

Click **Apply**.

Click **OK**.

The view title updates.

FLOOR PLAN
1/4″ = 1′-0″

15. Activate the Home ribbon.

In the Layers panel drop-down, select **Select Layer Standard**.

16.

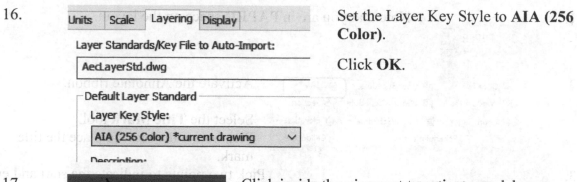

Set the Layer Key Style to **AIA (256 Color)**.

Click **OK**.

17.

Click inside the viewport to activate model space.

On the Status Bar: Set the Global Cut Plane to 3'6".

That way you see the doors and windows in the floor plan. You can only set the global cut plane in model space.

18.

Left click on the Global Cut Plane icon.

Set the Cut Height to **3'6"**.

Click **OK**.

19.

Verify that the **Add Annotative Scales** is enabled on the status bar.

20.

for use current <"use current">: A-Anno-Dims

Type **DIMLAYER**.

Type **A-Anno-Dims**.

This ensures any dimensions are placed on the A-Anno-Dims layer, regardless of which layer is current.

21.

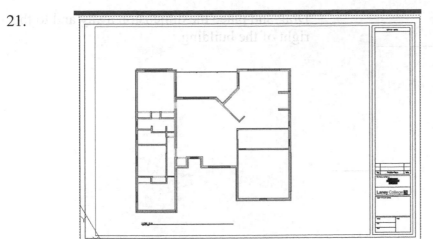

Zoom out to see how the layout sheet appears.

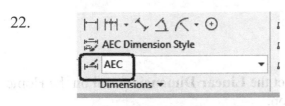

Activate MODEL space in the viewport on the A102 02 FLOOR PLAN layout tab.

22.

Set the Dimension Style to **AEC**.

23.

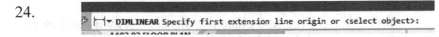

Select the **Linear Dimension** tool on the Home ribbon.

24.

`DIMLINEAR Specify first extension line origin or <select object>:`

Click **ENTER** to select an object.

25.

Select the right side of the building.

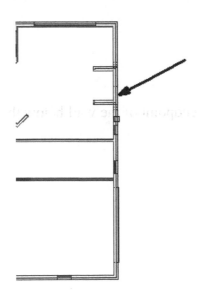

26.

Drag and place the dimension away and to the right of the building.

27.

Select the **Linear Dimension** tool on the Home ribbon.

28.

Select the upper right corner of the building.

Select the midpoint of the window.

Drag and place the dimension away and to the right of the building.

29.

Select **DIMCONTINUE**.

30.

Select the endpoint of the wall below the window.

Endpoint

31.
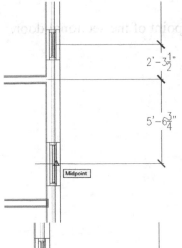
Select the midpoint of the wall below the door.

32.
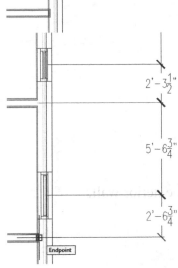
Select the endpoint of the wall below the door.

33.

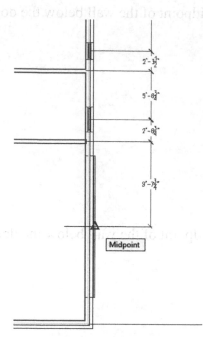

Select the midpoint of the sectional door.

34.

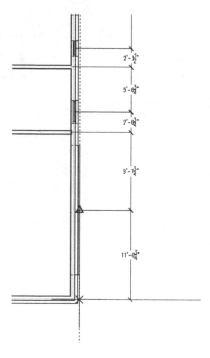

Select the south exterior wall.

35. 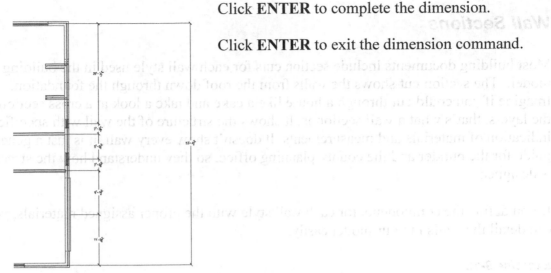 Click **ENTER** to complete the dimension.

Click **ENTER** to exit the dimension command.

36. 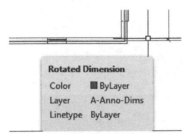 Hover over a dimension.

The dimension should automatically be placed on A-Anno-Dims because the DIMLAYER system variable was set to that layer.

37. Switch back to Paper space.

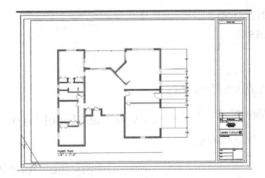

38. Save as *ex9-11.dwg*.

You can insert blocks into model or paper space. Note that you did not need to activate the viewport in order to place the Scale block.

Wall Sections

Most building documents include section cuts for each wall style used in the building model. The section cut shows the walls from the roof down through the foundation. Imagine if you could cut through a house like a cake and take a look at a cross section of the layers, that's what a wall section is. It shows the structure of the wall with specific indication of materials and measurements. It doesn't show every wall, it is just a general guide for the builder and the county planning office, so they understand how the structure is designed.

If you define the components for each wall style with the proper assigned materials, you can detail the walls in your model easily.

Exercise 9-12:

Create a Wall Section

Drawing Name: wall section.dwg
Estimated Time: 15 minutes

This exercise reinforces the following skills:

- ❑ Wall Section
- ❑ Layouts

1. Open *wall section.dwg.*

2. Activate the Model layout tab.

3. Apply the Floor Plan layer state.

   This layer state freezes several layers.

4. Activate the **Annotate** ribbon.

   Select the **Section (Sheet Tail)** tool.

5.

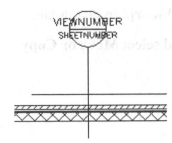

Draw the section through the south wall of the garage area.

Select a point below the wall.
Select a point above the wall.
Right click and select ENTER.
Left click to the right of the section line to indicate the section depth.

6.

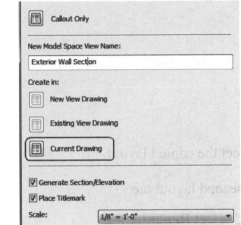

Type Exterior Wall Section in the View Name field.

Enable Current Drawing.

Enable Generate Section/Elevation.

Enable Place Title mark.
Set the Scale to 1/8″ = 1′-0″.

7.

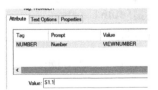

Place the view to the right of the South Elevation view.

8.

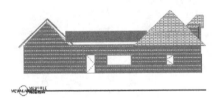

Double click on the view title for the wall section to bring up the attribute editor.

Type **1/8″ = 1′-0″** for the SCALE.

Type **Exterior Wall Section** for the TITLE.

Click ESC to release and update.

9.

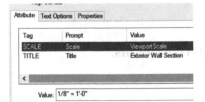

Select the View title bubble.

Type **S1.1** for the NUMBER.
Click **OK**.

Click **ESC** to release and update.

10. Highlight the **Floorplan** layout tab.

   Right click and select **Move or Copy**.

11. Enable **Create a copy**.

   Highlight **(move to end)**.

   Click **OK**.

12. Select the copied layout.

13. Highlight the second layout tab.

   Right click and select **Rename**.

14. Modify the layout name to **S1.2 Details**.

15. Unlock the viewport.
   Position the Exterior Wall Section in the viewport.

   Hint: Use the Named View that was created to adjust the view.

   Adjust the size of the viewport.

   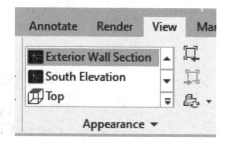

16.

Layer	Viewport
Display locked	Yes
Annotation scale	1/8" = 1'-0"
Standard scale	1/8" = 1'-0"
Custom scale	0"
Layer property o...	No
Visual style	2D Wireframe
Shade plot	Realistic

Select the viewport.
Set the Annotation scale to **1/8″ = 1′0″**.
Set the Standard scale to **1/8″ = 1′0″**.

Set the Shade plot to **Realistic**.

Lock the display.

The title bar will not display unless the view is set to the correct scale.

17. Click ESC to release the viewport.

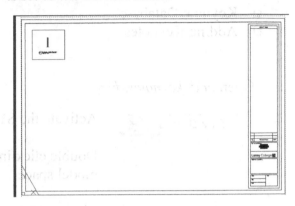

18. Save as *ex9-12.dwg*.

EXTRA: Create a section view for the interior wall.

Keynotes

Keynotes are used to standardize how materials are called out for architectural drawings. Most keynotes use a database. The database is created from the Construction Specifications Institute (CSI) Masterformat Numbers and Titles standard. ACA comes with several databases which can be used for keynoting or you can use your own custom database. The database ACA uses is in Microsoft Access format (*.mdb). If you wish to modify ACA's database, you will need Microsoft Access to edit the files.

Exercise 9-13:
Add Keynotes

Drawing Name: add_keynotes.dwg
Estimated Time: 25 minutes

This exercise reinforces the following skills:

- ❑ Keynote Database
- ❑ Adding Keynotes

1.   Open *add_keynotes.dwg*.

2.   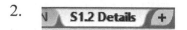   Activate the **S1.2 Details** layout tab.

Double click inside the wall section viewport to activate model space.

If you add the keynotes inside the viewport on the sheet, they will automatically use the annotative scale of that viewport and will be visible.

3.      *Verify that the keynote database is set properly in Options.*

Type **OPTIONS** to launch the Options dialog.

Select the AEC Content tab.

Select the Add/Remove button next to **Keynote Databases**.

4.

Path
. C:\ProgramData\Autodesk\ACA 2020\enu\Details\Details (UK)
. C:\ProgramData\Autodesk\ACA 2020\enu\Details\Details (Global)
C:\ProgramData\Autodesk\ACA 2020\enu\Styles\Imperial
C:\ProgramData\Autodesk\ACA 2020\enu\Styles\Metric
C:\ProgramData\Autodesk\ACA 2020\enu\Details\Details (US)

Verify that the paths and file names are correct.

Click **OK**.

Many companies set up their own keynote files.

Close the Options dialog.

5. Activate the Annotate ribbon.

Select the **Reference Keynote (Straight)** tool on the Keynoting panel.

6. Select the outside component of the wall section.

The Keynote database dialog window will open.

7. Select the *AecKeynotes (US).mdb*.

s\Details (US)\AecKeynotes (US).mdb

8. Locate the keynote for the **Clay/facing brickwork**.

04 21 00 - Clay Unit Masonry
04 21 00.A1 - Standard Brick - 3/8" Joint
04 21 00.A2 - Standard Brick - 1/2" Joint
04 21 00.A3 - Engineer Brick - 3/8" Joint
04 21 00.A4 - Engineer Brick - 1/2" Joint

Highlight and Click **OK**.

9. Select the outside component to start the leader line.
Left click to locate the text placement.
Click ENTER to complete the command.

10. Select the **Reference Keynote (Straight)** tool on the Keynoting panel.

Tail)   Reference Keynote (Straight)

11. Select the *AecKeynotes (US).mdb* database from the drop-down list.

s\Details (US)\AecKeynotes (US).mdb

12. Locate the air barrier keynote.

07 26 00 - Vapor Retarders
07 26 00.A1 - Moisture Barrier
07 26 00.A2 - Building Felt
07 26 00.A3 - 6 Mil Polyethylene
07 26 00.A4 - Vapor Retarder
07 26 00.A5 - Air Barrier

Highlight and Click **OK**.

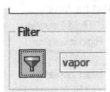

Filter   vapor

You can use the Filter tool at the bottom of the dialog to search for key words.

13.

Left click to place the leader and text.

14.

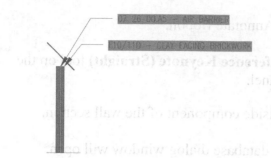

To turn off the shading on the keynotes, turn off the field display.

Type **FIELDDISPLAY** and then **0**.

15.

Select the **Reference Keynote (Straight)** tool on the Keynoting panel.

16.

Select the top of the wall.

In the Filter text field, type **insulation**.
Left click on the Filter icon.

17.

```
07 21 00 - Thermal Insulation
   07 21 00.A1 - R-11 Batt Insulation
   07 21 00.A2 - R-13 Batt Insulation
   07 21 00.A3 - R-15 Batt Insulation
   07 21 00.A4 - R-19 Batt Insulation
   07 21 00.A5 - R-21 Batt Insulation
   07 21 00.A6 - R-22 Batt Insulation
   07 21 00.A7 - R-25 Batt Insulation
   07 21 00.A8 - R-30 Batt Insulation
   07 21 00.A9 - R-38 Batt Insulation
   07 21 00.A10 - Batt Insulation
   07 21 00.B1 - 1/2" Rigid Insulation
   07 21 00.B2 - 1" Rigid Insulation
   07 21 00.B3 - 1 1/2" Rigid Insulation
   07 21 00.B4 - 2" Rigid Insulation
   07 21 00.B5 - 2-1/2" Rigid Insulation
   07 21 00.B6 - 3" Rigid Insulation
```

Locate the keynote for the rigid insulation and add to the view.

Note that if you select the component on the wall section, ACA will automatically take you to the correct area of the database to locate the proper keynote.

18.

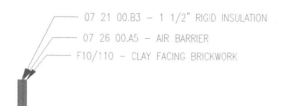

Left click to place the leader and the text.

19.

Select the **Reference Keynote (Straight)** tool on the Keynoting panel.

20.

Select the top of the wall.

In the Filter text field, type s**tud**.
Left click on the Filter icon.

21.

Details\Details (US)\AecKeynotes (US).mdb

Select the *AecKeynotes (US).mdb.*

22.

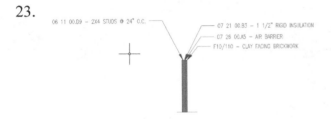

Locate the keynote for the 2x4 Studs.

- Division 06 - Wood, Plastics, and Composites
  - 06 11 00 - Wood Framing
    - 06 11 00.A3 - Existing Studs
    - 06 11 00.D7 - 2x4 Studs
    - 06 11 00.D8 - 2x4 Studs @ 16" O.C.
    - 06 11 00.D9 - 2x4 Studs @ 24" O.C.
    - 06 11 00.F6 - 2x6 Studs
    - 06 11 00.F7 - 2x6 Studs @ 16" O.C.
    - 06 11 00.F8 - 2x6 Studs @ 24" O.C.

23.

Left click to place the leader and the text.

24.

Select the **Reference Keynote (Straight)** tool on the Keynoting panel.

25.

Select the top of the wall.

In the Filter text field, type gypsum.
Left click on the Filter icon.

26.

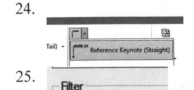

- 09 29 00.C1 - 1/2" Gypsum Wallboard
- 09 29 00.C2 - 1/2" Type "X" Gypsum Wallboard
- 09 29 00.C3 - 1/2" M.R. Gypsum Board
- 09 29 00.D1 - 5/8" Gypsum Wallboard
- 09 29 00.D2 - 1 Layer 5/8" Gypsum Board
- 09 29 00.D3 - 1 Layer 5/8" Gypsum Board On Each Side
- 09 29 00.D4 - Add 1 Layer 5/8" Gypsum Board To Existing Wall

Locate the keynote for the **1 Layer 5/8" Gypsum Board.**

Click **OK.**

27.

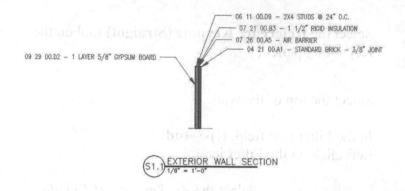

06 11 00.D9 – 2X4 STUDS @ 24" O.C.
07 21 00.B3 – 1 1/2" RIGID INSULATION
07 26 00.A5 – AIR BARRIER
04 21 00.A1 – STANDARD BRICK – 3/8" JOINT

09 29 00.D2 – 1 LAYER 5/8" GYPSUM BOARD

S1.1 EXTERIOR WALL SECTION
1/8" = 1'–0"

Save as *ex9-13.dwg*.

EXTRA: If you created a section view of the interior wall, add keynotes to that section view.

Exercise 9-14:

Add Keynote Legend

Drawing Name: keynote_legend.dwg
Estimated Time: 25 minutes

This exercise reinforces the following skills:

- ❑ Keynote Legend
- ❑ Keynote display
- ❑ View Manager
- ❑ Layouts

1.   Open *keynote_legend.dwg*.

2.   **Model**   Activate the Model layout tab.

3.   Annotate   Render   View

      Exterior Wall Se
      South Elevation
      Top
      Appearance

      Switch to the View ribbon.

      Activate the **Exterior Wall Section** named view.

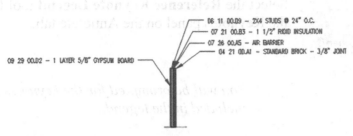

Notice the keynote numbers and text.

4.

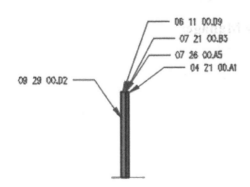

Switch to the Annotate ribbon.

Locate the **Reference Keynote Display** tool and select it.

5.

Enable **Reference Keynote – Key only**.

Enable **Uppercase**.

Click **OK**.

The keynotes update to display keys only.

6. Select the **Reference Keynote Legend** tool from the Keynotes panel on the Annotate tab.

7.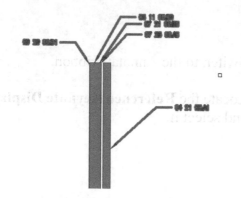

You will be prompted for the keynotes to be included in the legend.

Window around to select all the keynotes.

Click ENTER to complete the selection.

8.

REFERENCE KEYNOTES

DIVISION 04 - MASONRY		
04 21 00 - CLAY UNIT MASONRY		
04 21 00.A1	–	STANDARD BRICK – 3/8" JOINT
DIVISION 06 - WOOD, PLASTICS, AND COMPOSITES		
06 11 00 - WOOD FRAMING		
06 11 00.09	–	2X4 STUDS @ 24" O.C.
DIVISION 07 - THERMAL AND MOISTURE PROTECTION		
07 21 00 - THERMAL INSULATION		
07 21 00.B3	–	1 1/2" RIGID INSULATION
07 26 00 - VAPOR RETARDERS		
07 26 00.A5	–	AIR BARRIER
DIVISION 09 - FINISHES		
09 29 00 - GYPSUM BOARD		
09 29 00.D1	–	5/8" GYPSUM WALLBOARD

Click to place the keynote legend.

There will be a slight pause before the legend is placed while ACA compiles the data.

9. Right
Front
Back
SW Isometric
SE Isometric
NE Isometric
View Manager...

Switch to the View tab.

Launch the **View Manager**.

10.

Set Current

New...

Update Layers

Edit Boundaries...

Delete

Click **New**.

11.

View name:	Exterior Wall Keynotes Legend
View category:	<None>
View type:	Still

Boundary

○ Current display
◉ Define window

Type **Exterior Wall Keynotes Legend**.

Enable **Define Window**.

Click the window icon.

Window around the keynotes legend.

Click **ENTER**.

Click **OK**.
The new named view is listed.

12.

Model Views
  Door Legend
  Door Schedule
  Exterior Wall Keynotes
  Exterior Wall Section
  HD1
  Master Bath Elevation
  Single Interior Door
  South Elevation
Layout Views
Preset Views

Name	Exterior Wall Key...
Category	<None>
UCS	World
Layer snapshot	Yes
Annotation s...	1:25
Visual Style	2D Wireframe
Background...	<None>
Live Section	<None>

Animation

| View type | Still |
| Transition type | Fade from black... |

OK   Cancel   Apply   Help

You can see a preview in the preview window.

Click **OK**.

13.

Type a commana

ting   **S1.2 Details**   +

Select the **S1.2 Details** layout.

14.   **Model**   Activate the Model layout tab.

15.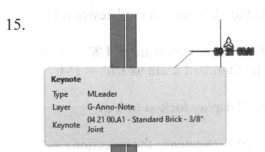

Keynote

Type   MLeader
Layer   G-Anno-Note
Keynote 04 21 00.A1 - Standard Brick - 3/8"
         Joint

If you hover over one of the keynotes, you will see only one annotative scale.

You also will see the keynote information and the assigned layer.

16.   g   **S1.2 Details**   Select the **S1.2 Details** layout.

17.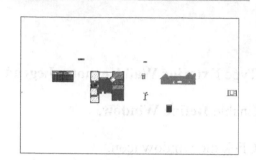

Switch to the Layout ribbon.

Select **Rectangular** viewport.

18.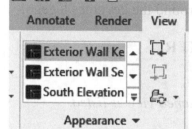

Draw a rectangular viewport on the layout sheet.

Double click inside the viewport to activate model space.

19.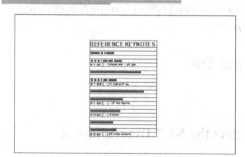

Switch to the View ribbon.

Activate the **Exterior Wall Keynotes Legend** named view.

20.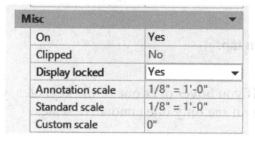

The view in the viewport will adjust to display the keynote legend.

Select the viewport.

Right click and select **Properties**.

21.

Misc	▼
On	Yes
Clipped	No
Display locked	Yes ▼
Annotation scale	1/8" = 1'-0"
Standard scale	1/8" = 1'-0"
Custom scale	0"

General	
Color	■ ByLayer
Layer	Viewport

Place the viewport on the Viewport layer.

Set the Annotation scale to **1/8" = 1'-0"**.
Set the Standard scale to **1/8" = 1'-0"**.

Set the Display locked to **Yes**.

Click ESC to release the selection.

22.

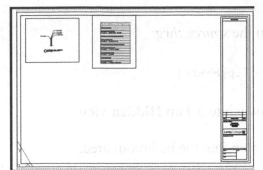

Save as *ex9-14.dwg*.

Exercise 9-15:

Copy and Paste between Drawings

Drawing Name: target.dwg, source.dwg
Estimated Time: 10 minutes

This exercise reinforces the following skills:

- ❑ Copy
- ❑ Paste

1. Open *target.dwg*.
   Open *source.dwg*.

2.

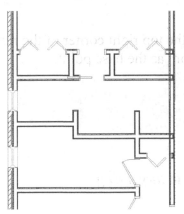

In the *target.dwg*:

Open the Floorplan layout tab.

Double left click inside the viewport to activate MODEL space.

Zoom into the bathroom area.

3.

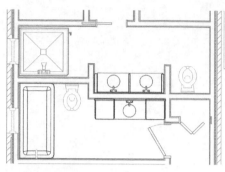

In the *source.dwg*:

[−][Top][Hidden]

Switch to a **Top Hidden** view.

Zoom into the bathroom area.

4.

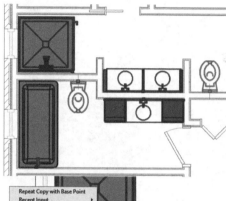

Select the blocks located in the bathrooms.

5.

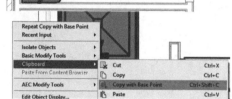

Right click and select **Clipboard→Copy with Base Point**.

6.

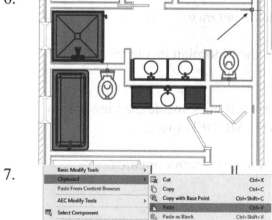

Select the top right corner of the master bathroom as the base point.

7.

Switch to *target.dwg*.

8.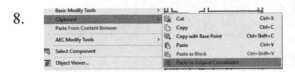

Right click and select **Copy→Paste to Original Coordinates**.

Paste to Original Coordinates only works if you are using similar floorplans/coordinate systems. If there is a difference between the two drawings, use Copy with Base Point and then select an insertion point in the target drawing.

Save as *ex9-15.dwg*.

9.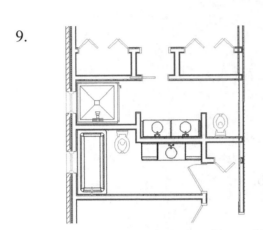

Section Views

Section views show a view of a structure or building as though it has been sliced in half or along an imaginary plane. This is useful because it gives a view through spaces and can reveal the relationships between different parts of the building which may not be apparent on the plan views. Plan views are also a type of section view, but they cut through the building on a horizontal plane instead of a vertical plane.

Exercise 9-16:

Create a Section View

Drawing Name: section.dwg
Estimated Time: 20 minutes

This exercise reinforces the following skills:

- ❑ Sections
- ❑ Elevations
- ❑ Layouts
- ❑ Viewports
- ❑ View Titles

1. Open *section.dwg*.

2. Activate the **Model** layout tab.

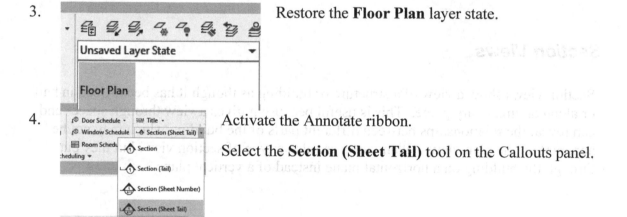

3. Restore the **Floor Plan** layer state.

4. Activate the Annotate ribbon.

   Select the **Section (Sheet Tail)** tool on the Callouts panel.

5. 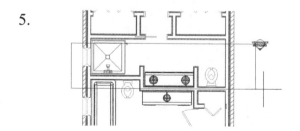 Draw a horizontal line through the master bathroom walls as shown.

6.

Name the View **Master Bath Elevation**.
Enable **Generate Section/Elevation**.
Enable **Place Titlemark**.
Set the Scale to **1/2″ = 1′-0″**.

Click **Current Drawing**.

Place the view in the current drawing to the right of the building.

7.

Right click on the + layout tab.

Select **From Template**.

8.

Select the **Arch_D** template from the classwork folder.

Click **Open**.

9.

Select the **D-size Layout**.

Click **OK**.

10.

Highlight the layout tab.

Right click and select **Rename**.

11.

Rename the layout name to **Master Bath Elevation**.

12.

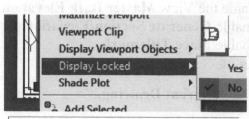

Unlock the viewport display.
Select the Viewport, right click and set Display Locked to No.

13.

Activate Model space by double clicking inside the viewport.

14.

Switch to the View ribbon.
Double left click on the **Master Bath Elevation** view.

Position and resize the viewport to show the Master Bath Elevation.

15.

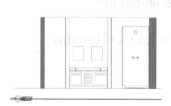

Select the viewport and set the scales to ½" = 1'-0".

Hint: You will not see the title bar unless the annotation scale matches the viewport scale and is set correctly. Remember this is the scale you selected when you created the view.

Set the Display Locked to **Yes**.

16.

Return to the Model space by clicking inside the viewport
.

Zoom into the Master Bath Elevation.

17.

Change the values of the view title attributes as shown.

18.  Adjust the position of the view title to be below the elevation.

19.  Adjust the viewport so the view title is visible using the grips.

20. Save as *ex9-16.dwg*.

To create an elevation view with different visual styles, create a camera view.

Exercise 9-17:

Add a Door Elevation

Drawing Name: door elevation.dwg
Estimated Time: 20 minutes

This exercise reinforces the following skills:

- ❑ Elevations
- ❑ View Titles
- ❑ View Manager
- ❑ Named Views

1.   Open *door elevation.dwg*.

2.   Activate the **Model** tab.

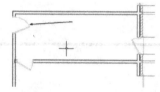

3.

Zoom into the laundry room area.

We will create an elevation for the interior door frame.

4.

Activate the Annotate ribbon.

Select the **Elevation (Sheet No)** tool.

5.

Left click to the left of the door to place the elevation marker.

Orient the marker so it is pointing toward the door.

6.

Name the view **Single Interior Door**.
Enable **Generate Section/Elevation**.
Enable **Place Titlemark**.
Set the Scale to **1/2″ = 1′-0″**.

Click **Current Drawing**.

7.

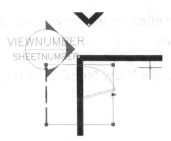

Draw a rectangle around the door to define the scope of the elevation.

8.

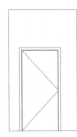

Left click to the right of the floor plan to place the elevation view.

9.

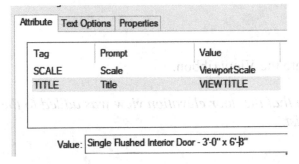

Double click on the title attribute.

Change the title to **Single Flushed Interior Door - 3'-0" x 6'-8"**.

10.

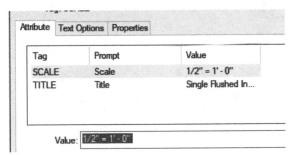

Change the Scale to ½" = 1'-0".

Click **OK.**

11.

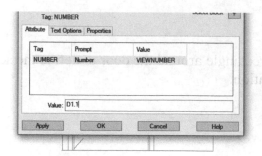

Double click on the view number attribute.

Change the View Number to **D1.1**.

I click to the right of the floor plan to place the elevation view.

This is what the view should look like.

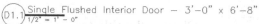

12.

Activate the View ribbon.

Notice that the door elevation view was added to the view list.

13. Save as *ex9-17.dwg*.

Exercise 9-18:

Add a Door Details Layout

Drawing Name: door detail.dwg
Estimated Time: 45 minutes

This exercise reinforces the following skills:

- ❑ Layouts
- ❑ Named Views
- ❑ Annotation Scales

1. Open *door detail.dwg.*

2. 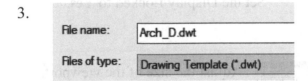 Right click on the + add layout tab.
    Select **From Template**.

3. 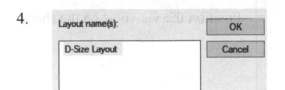 Select the **Arch_D** template from the classwork folder.

    Click **Open**.

4. Select the **D-size Layout**.

    Click **OK**.

5. 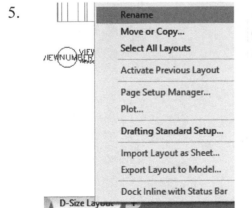 Highlight the layout tab.

    Right click and select **Rename**.

6. **E1.2 Door Details**

    Rename the layout name to **E1.2 Door Details**.

    Activate this layout.

7.

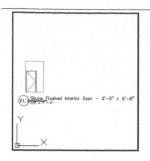

Resize the viewport.

Unlock the display.

Double click inside the viewport to activate MODEL space.

8.

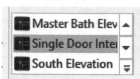

Activate the View ribbon.

Select the **Single Door Interior** view.

Position inside the viewport.

9.

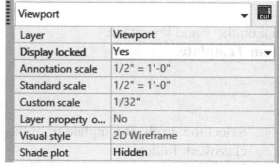

Set the Viewport scale to **1/2" = 1'-0"**.
Set the Annotation scale to **1/2" = 1'-0"**.

Set the Shade Plot to **Hidden**.

Set the Display Locked to **Yes**.

10.

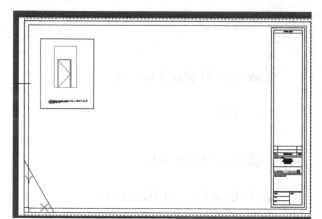

Adjust the size of the viewport.

Position the viewport on the sheet.

11.

Save as *ex9-18.dwg*.

Callouts

Callouts are typically used to reference drawing details. For example, builders will often reference a door or window call-out which provides information on how to frame and install the door or window. Most architectural firms have a detail library they use to create these callouts as they are the same across building projects.

Exercise 9-19:

Add a Callout

Drawing Name: callout.dwg
Estimated Time: 45 minutes

This exercise reinforces the following skills:

- ❑ Callouts
- ❑ Viewports
- ❑ Detail Elements
- ❑ Layouts
- ❑ Hatching

1. Open *callout.dwg*.

2.  Activate **Model** space.

3. Zoom into the door elevation view.

    Hint: Use the View Manager to locate the view.

    We will create a callout for the interior door frame.

4. 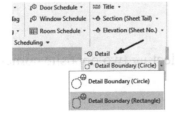 Activate the Annotate ribbon.

    Turn ORTHO **ON**.

    Locate and select the **Detail Boundary (Rectangle)** tool on the Detail drop-down list on the Callouts panel.

5.

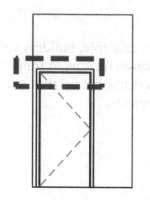

Draw a small rectangle to designate the callout location around the top of the door.

Hint: Turn off OSNAPs to make it easier to place the rectangle.

Select the midpoint of the left side of the rectangle for the start point leader line.
Left click to the side of the callout to place the label.

Click **ENTER** or right click to accept.

6.

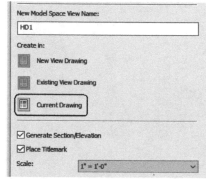

Type **HD1** in the View Name field.

Enable **Generate Section/Elevation**.

Enable **Place Titlemark**.

Set the Scale to **1″ = 1′-0″**.

Click on the **Current Drawing** button.

7.

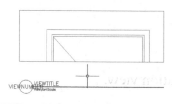

Place the view to the right of the elevation.

8.

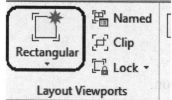

Activate the **S1.2 Details** layout tab.

9.

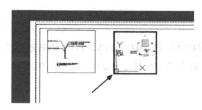

Activate the Layout ribbon.

Select the **Rectangular** viewport tool.

10.

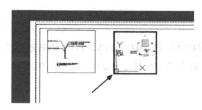

Draw a rectangle by selecting two opposite corners to place the viewport.

11.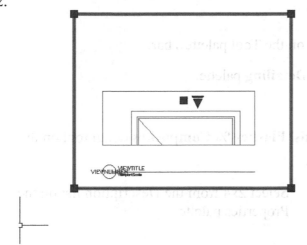

Set the Viewport scale to **1" = 1'-0".**
Set the Annotation scale to **1" = 1'-0".**
Set the Shade Plot to **Hidden.**

12.

Adjust the size of the viewport.

Position the door view in the viewport.

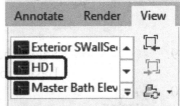

Hint: Use the Named View list to bring up the view.

Re-set the scale in Properties.
Lock the display.

13.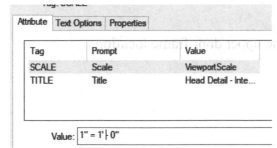

Place the viewport on the viewport layer.

14.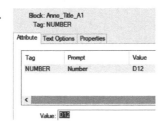

Double click inside the viewport to activate model space.

Select the view title for the door callout.

Type **1" = 1'-0"** for the SCALE.

Type **Head Detail – Interior Door** for the TITLE.

15.

Select the View title bubble.

Type **HD1** for the NUMBER.

Click **ESC** to release and update.

16. 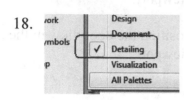 Use the grips to position the text and the title line so it looks correct.

17. 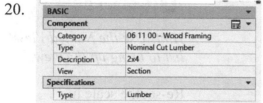 Launch the **Document** tool palette from the Annotate ribbon.

18. Right click on the Tool palette's bar.

Enable the **Detailing** palette.

19. Select the **06- Woods, Plastics % Composites** detail tool on the Basic tab.

20. Select 2x4 from the Description list on the Properties palette.

21. Right click and select Rotate to rotate the detail.

22. 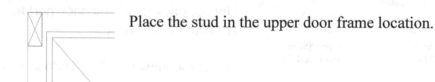 Place the stud in the upper door frame location.

23.

Geometry	
Position X	394'-9 25/32"
Position Y	20'-6 7/8"
Position Z	0"
Scale X	-1.30000
Scale Y	3.50000
Scale Z	1.00000

Select the stud detail.

Right click and select Properties.
Change the X scale to **1.3**.
Change the Y scale to **3.5**.

The stud size will adjust to fit the frame.

To get those values, I took the desired dimension 12" and divided it by the current dimension 3.5 to get 3.5. I took the desired dimension of 2.0 and divided it by the current dimension of 1.5 to get 1.3. To determine the dimensions, simply measure the current length and width and the desired length and width.

24.

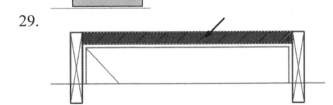

Copy and place the stud to the other side of the door.

25.

Activate the Home ribbon.

Set the layer **A-Detl-Wide** current.

26. Draw a rectangle in the header area.

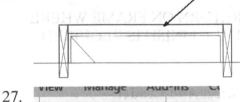

27.

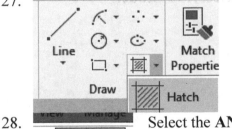

Select the **Hatch** tool.

28. Select the **ANSI31** hatch.

ANSI31

29.

Place a hatch inside the rectangle.

30.

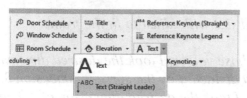

Select **Close Hatch Creation** on the ribbon.

31.

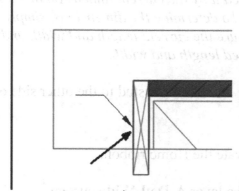

Activate the Annotate ribbon.

Select the **Text (Straight Leader)** tool.

32.

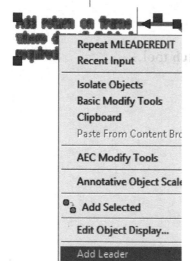

Select the left side of the stud as the start for the leader.

Click **ENTER** when prompted to Set the text width.

33.

ADD RETURN ON FRAME
WHERE DRYWALL FINISH IS REQUIRED

Text should read:

ADD RETURN ON FRAME WHERE DRYWALL FINISH IS REQUIRED

34.

Repeat MLEADEREDIT
Recent Input

Isolate Objects
Basic Modify Tools
Clipboard
Paste From Content Bro

AEC Modify Tools

Annotative Object Scale

Add Selected

Edit Object Display...

Add Leader
Remove Leader

Select the text.
Right click and select **Add Leader**.

35.

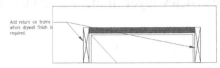

Add a second leader to the other side of the door frame.

Click **ESC** to release the selection.

36.

Hover over the leader that was just placed.

Notice that it is on the **G-Anno-Note** layer. *This layer was created automatically when you placed the leader.*

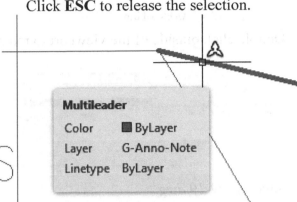

37. ⊢⊣ Select the Linear Dimension tool on the Annotate ribbon.

38. _____ Place a dimension in the opening.

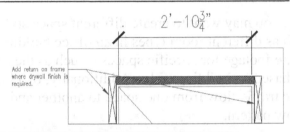

39.

Text	
Fill color	None
Fractional type	Horizontal
Text color	■ ByBlock
Text height	3/32"
Text offset	1/8"
Text outside align	Off

Select the dimension.
Right click and select **Properties**.

Change the Text height to **3/32"**.

40.

Text rotation	0.00
Text view direction	Left-to-Right
Measurement	2'-10 3/4"
Text override	THROAT OPENING = PARTITION WIDTH + 1/8" MIN.

Modify the text to read:

THROAT OPENING = PARTITION WIDTH + 1/8" MIN.

41. 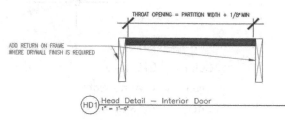 Freeze the **G-Elev** layer in the active viewport.

42. Double click outside of the viewport to return to paper space.

This is what the view should look like.

Head Detail — Interior Door

43. Save as *ex9-19.dwg*.

Space Styles

Depending on the scope of the drawing, you may want to create different space styles to represent different types of spaces, such as different room types in an office building. Many projects require a minimum square footage for specific spaces – such as the size of bedrooms, closets, or common areas – based on building code. Additionally, by placing spaces in a building, you can look at the traffic flow from one space to another and determine if the building has a good flow pattern.

You can use styles for controlling the following aspects of spaces:

• Boundary offsets: You can specify the distance that a space's net, usable, and gross boundaries will be offset from its base boundary. Each boundary has its own display components that you can set according to your needs.

• Name lists: You can select a list of allowed names for spaces of a particular style. This helps you to maintain consistent naming schemes across a building project.

• Target dimensions: You can define a target area, length, and width for spaces inserted with a specific style. This is helpful when you have upper and lower space limits for a type of room that you want to insert.

• Displaying different space types: You can draw construction spaces, demolition spaces, and traffic spaces with different display properties. For example, you might draw all construction areas in green and hatched, and the traffic areas in blue with a solid fill.

• Displaying different decomposition methods: You can specify how spaces are decomposed (trapezoid or triangular). If you are not working with space decomposition extensively, you will probably set it up in the drawing default.

Space Styles are created, modified, or deleted inside the Style Manager.

Exercise 9-20:
Creating Space Styles

Drawing Name: spaces.dwg
Estimated Time: 10 minutes

This exercise reinforces the following skills:

- Style Manager
- Space Styles

1. Open *spaces.dwg*.

2. Save as *ex9-20.dwg*.

3.  Activate the **Manage** ribbon.

   Select the **Style Manager**.

4. 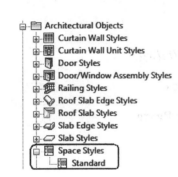 Locate the *Space Styles* inside the drawing.

   Note that there is only one space style currently available in the spaces.dwg.

5.  Select the **Open Drawing** tool.

   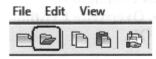

6.  Browse to the downloaded exercise folders.
   Change the files of type to **Drawing (*.dwg)**.
   Open the *space styles.dwg*.

7.

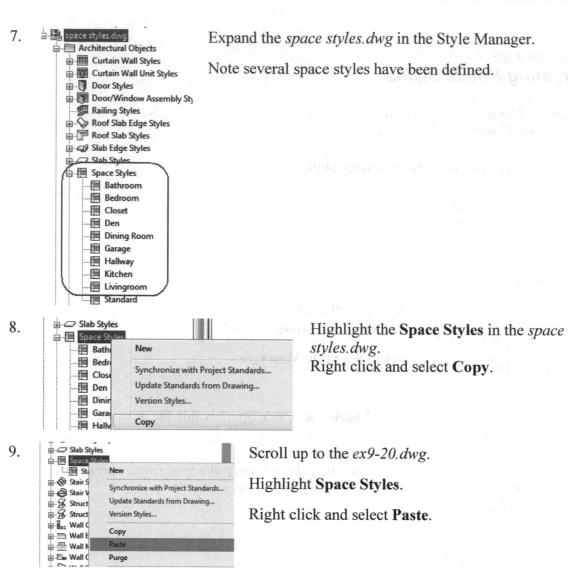

Expand the *space styles.dwg* in the Style Manager.

Note several space styles have been defined.

8.

Highlight the **Space Styles** in the *space styles.dwg*.
Right click and select **Copy**.

9.

Scroll up to the *ex9-20.dwg*.

Highlight **Space Styles**.

Right click and select **Paste**.

10.

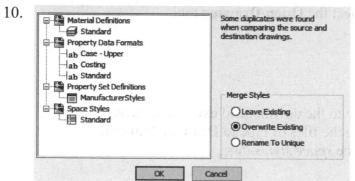

If this dialog appears advising of duplicate definitions,
enable **Overwrite Existing**.
Click **OK**.

11. 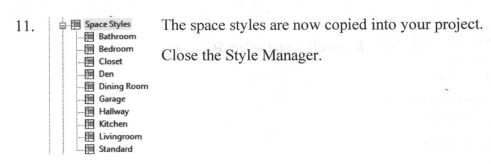 The space styles are now copied into your project.

Close the Style Manager.

12. Save as *ex9-20.dwg*.

Exercise 9-21:

Adding Spaces

Drawing Name: add_space.dwg
Estimated Time: 15 minutes

This exercise reinforces the following skills:

☐ Layouts
☐ Spaces

1. Open *add_space.dwg*.

2. 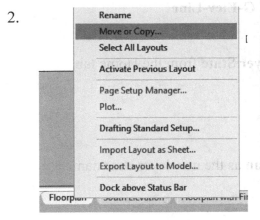 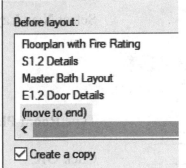 Right click on the Floorplan layout tab and select **Move or Copy**.

Enable **Create a copy**.

Highlight **(move to end)**.

Click **OK**.

3. 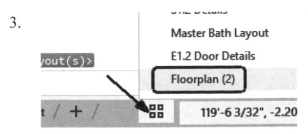 Select the copied layout.

You may need to click on the layouts icon in the Status Bar to locate the copied layout.

9-85

4.

Rename	
Move or Copy...	
Select All Layouts	
Activate Previous Layout	
Activate Model Tab	
Page Setup Manager...	
Plot...	
Drafting Standard Setup...	
Import Layout as Sheet...	
Export Layout to Model...	
Dock above Status Bar	

Right click and select **Rename**.

5. etails ⟩ A102 03 ROOM PLAN ╱   Modify the layout name to **A102 03 ROOM PLAN**.

6.
💡 ☀ 📋 🔓 ⬛ G-Elev-Iden
💡 ☀ 📋 🔓 ⬛ G-Elev-Line
💡 ☀ 📋 🔓 ⬛ G-Sect-Iden
💡 ☀ 📋 🔓 ⬛ G-Sect-Line

Double click inside the viewport to activate model space.

Freeze in Viewport the following layers:

G-Sect-Iden
G-Sect-Line
G-Elev-Line

7.

Unsaved Layer State ▾
Floor Plan
New Layer State...
Manage Layer States...

Select **New Layer State** from the Home tab.

8.

New layer state name:

Room Plan

Type **Room Plan** as the new layer state name.

Click **OK**.

9.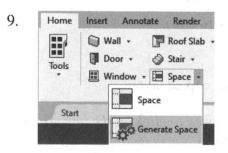

Select the **Generate Space** tool from the Home ribbon on the Build panel.

10. 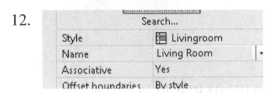 On the Properties palette, select the space style to apply.

Left click inside the designated space to place the space.

11. 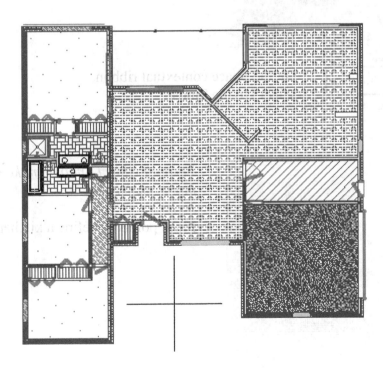 If you hover over a space, you should see a tool tip with the space style name.

12. Select a space.

In the Properties palette select the Name for the SPACE using the drop-down list.

This is the name that will appear on the room tag.

13. Repeat for all the spaces.
Click ESC to release the selection of each space before moving on to the next space.

Verify that only one space is selected to be renamed each time or you may assign a name to the wrong space unintentionally.

14. Save as *ex9-21.dwg*.

Exercise 9-22:
Modifying Spaces

Drawing Name: space_modify.dwg
Estimated Time: 15 minutes

This exercise reinforces the following skills:

    ❑   Spaces

1.   Open *space_modify.dwg*.

2.   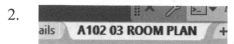   Open A102 03 ROOM PLAN layout.

3.   Double click inside the viewport to activate MODEL space.

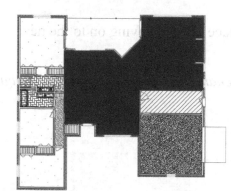

Select the space located in the living room area.

4.      On the Space contextual ribbon, select **Divide Space**.

5.   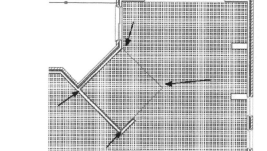   *Turn off Object Snap to make selecting the points easier.*

Place points to define a kitchen area.

6.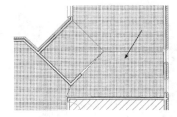

Repeat the Divide Space command to place lines to define the den and dining room areas.

Make sure each divided space is a closed polygon.

7.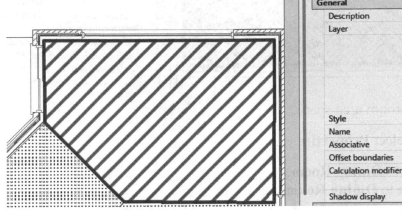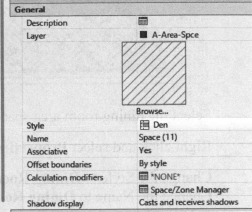

Select the den area space.

Right click and select **Properties**.

Change the Style to **Den**.
Change the Name to **Den.**

Click **ESC** to release the selection.

8.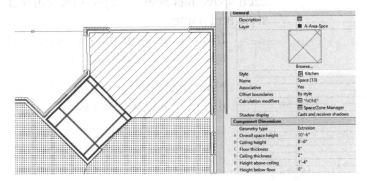

Select the kitchen area space.

Right click and select **Properties**.

Change the Style to **Kitchen**.
Change the Name to **Kitchen**.

Click **ESC** to release the selection.

9.

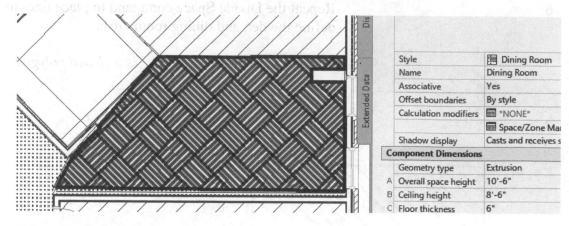

	Style	🔳 Dining Room
	Name	Dining Room
	Associative	Yes
	Offset boundaries	By style
	Calculation modifiers	🔳 *NONE*
		🔳 Space/Zone Ma...
	Shadow display	Casts and receives s...
Component Dimensions		
	Geometry type	Extrusion
A	Overall space height	10'-6"
B	Ceiling height	8'-6"
C	Floor thickness	6"

Select the dining room area space.

Right click and select **Properties**.

Change the Style to **Dining Room**.
Change the Name to **Dining Room**.

Click **ESC** to release the selection.

10.

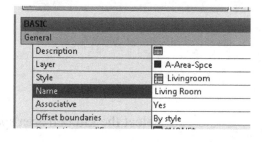

BASIC		
General		
Description	🔳	
Layer	■ A-Area-Spce	
Style	🔳 Livingroom	
Name	Living Room	
Associative	Yes	
Offset boundaries	By style	

Select the Living room space.
Right click and select Properties.
On the Design tab,
change the name to **Living Room**.

Repeat for the remaining spaces. Verify that each space has been assigned the correct style and name.

11.

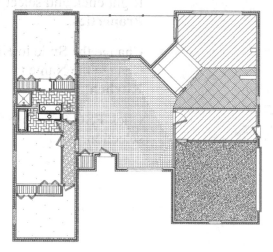

Inspect the floor plan to verify that all spaces have been placed.

Note that different space styles use different hatch patterns.

12. Save as *ex9-22.dwg*.

Property Set Definitions

Property set definitions are used to assign properties to AEC objects. These properties can then be used in tags and schedules.

Exercise 9-23:

Create a Room Tag

Drawing Name: schedule tag updated.dwg
Estimated Time: 45 minutes

This exercise reinforces the following skills:

- ❑ Spaces
- ❑ Property Set Definitions
- ❑ Style Manager
- ❑ Property Set Formats
- ❑ Attributes
- ❑ Add/Delete Scales
- ❑ Tags
- ❑ Palettes
- ❑ Blocks

1.   Open *schedule tag updated.dwg*.

2. | 1'-0" = 1'-0" ▾ | ⚙ ▾ | ✛ | Medium Detail ▾ |

Click on the **Model** tab.
Verify that the scale is set to **1'-0"-1'-0"**.
Set the Display to **Medium Detail**.

3.

Launch the Document tools palette.

Click on the Tags tab.

Locate the **Room Tag**.

Place on the top space in the drawing.

Click **ENTER** to accept placing in the center of the space.

4.  The available properties will display.

Click **OK**.

Repeat to tag the lower space.

5. 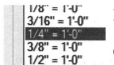 If you zoom in:

The tag is quite small.
It is using millimeter units.

Select the tag.

6.  On the Annotate ribbon:

Click **Add/Delete Scales**.

7. 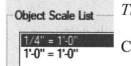 Select the top room tag.
Click **ENTER**.

Click **Add**.

8.  Highlight **¼" = 1'-0"**.

Click **OK**.

9. 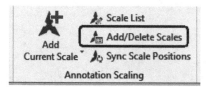 *There are now two annotative scales available for the tag.*

Click **OK**.

10. | Zoom out.

3/16" = 1'-0"
1/4" = 1'-0"
3/8" = 1'-0"
1/2" = 1'-0"
3/4" = 1'-0"
1" = 1'-0"
1 1/2" = 1'-0"
3" = 1'-0"
6" = 1'-0"
✓ 1'-0" = 1'-0"
Custom...
Xref scales
Percentages

1'-0" = 1'-0"

Set the view scale to ¼" = 1'-0".

11.

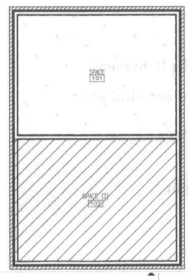

Both space tags now look OK.

We are going to create a space tag using imperial units.

Select the two room tags that were placed and delete.

12.

1'-0" = 1'-0" ▼   ☼ ▼   + Medium Detail ▼

Set the scale to **1'-0"-1'-0"**.
Set the Display to **Medium Detail**.

13.

Style Manager   Display Manager   Renovation Mode

Style & Display ▼

Switch to the Manage ribbon.

Click on **Style Manager**.

14.

- Documentation Objects
  - 2D Section/Elevation Styles
  - AEC Dimension Styles
  - Calculation Modifier Styles
  - Display Theme Styles
  - Property Data Formats
  - Property Set Definitions
    - DoorStyles
    - FrameStyles
    - GeoObjects
    - ManufacturerStyles
    - RoomFinishObjects
    - SpaceEngineeringObjects
    - SpaceObjects
    - SpaceStyles
    - ThermalProperties
    - WallStyles
    - WindowStyles
    - ZoneEngineeringObjects
  - Schedule Table Styles
  - Zone Styles
  - Zone Templates

Expand the **Documentation Objects** folder.

Expand **Property Set Definitions**.

Click on **SpaceObjects**.

| General | Applies To | Definition |

Select the **Definition** tab.

15. Click on **Add Automatic Property Set Definitions**.

16.

| Alphabetical | Categorized |

- Base Area ☑
- Base Area Minus Int... ☐
- Base Ceiling Area ☐

Click on the Alphabetical tab.

Enable **Base Area**.

17.

- Hyperlink ☐
- Layer ☐
- Length ☑
- Linetype ☐
- Maximum Area ☐

Enable **Length**.

18.

Enable **Name** and **Style**.

19.

Enable **Width**.

Click **OK**.

20.

You can see a list of properties which will be applied to space objects.

Click **Apply** to save your changes.

21.

Set the Format for the Property Set Definitions.

Highlight BaseArea.

In the Format column: use the drop-down list to select **Area**.

22.

Set BaseArea to **Area**.
Set Length to **Length – Short**.
Set Name to **Case – Sentence**.
Set Style to **Case – Sentence**
Set Width to **Length – Short**.
Click **Apply**.

23.

```
☐ lab Property Data Formats
    ── lab Area
    ── lab Case - Sentence
    ── lab Case - Upper
    ── lab Costing
    ── lab Length - Long
    ── lab Length - Nominal
    ── lab Length - Short
    ── lab Number - Object
    ── lab Standard
    ── lab Unit Suffix – Flow Cubic feet per minu
    ── lab Unit Suffix – Power Watts
    ── lab Unit Suffix – Temperature Fahrenheit
    ── lab Volume
```

Expand the **Property Data Formats** category.

24.

Real Numbers

Unit Type:	Area
Units:	Square feet
Unit Format:	Decimal
Precision:	0.0
Fraction Format:	Not Stacked

Highlight **Area** in the list.

General | Formatting

Click on the **Formatting** tab.

Set the Precision to **0.0**.

Verify that the other fields are set correctly.

25.

General | Formatting

General

Prefix:	
Suffix:	SF
Undefined:	?
Not Applicable:	NA

Verify that the Suffix is set to **SF**.

Click **Apply**.

26.

Real Numbers

Unit Type:	Length
Units:	Inches
Unit Format:	Architectural
Precision:	0'-0 1/4"
Fraction Format:	Not Stacked

Highlight Length – Short.

Set the Precision to ¼".

Click **Apply**.

Click **OK** to close the Style Manager.

27.

Unsaved Layer State

G-Area-Spce-Iden

Layers ▾

Set the **G-Area-Spce-Iden** layer current.

28. Switch to the Insert ribbon.

Click **Define Attributes**.

29.  In the Tag field:
Type **SPACEOBJECTS:STYLE**.
In the Prompt field:
Type **Style**.
In the Default field:
Type **Living Room**.
Set the Justification to **Center**.
Set the Text style to **RomanS**.
Disable **Annotative**.
Set the Text height to **1"**
Enable **Specify on-screen** for the insertion point.
Click **OK**.

30. Click to place the attribute in the drawing.
SPACEOBJECTS:STYLE Zoom into the attribute.

31. Click **Define Attributes**.

32. 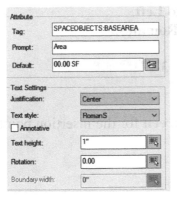 In the Tag field:
Type **SPACEOBJECTS:BASEAREA**.
In the Prompt field:
Type **Area**.
In the Default field:
Type **00.0 SF**.
Set the Justification to **Center**.
Set the Text style to **RomanS**.
Disable **Annotative**.
Set the Text height to **1"**

33.

Enable **Align below previous attribute definition** for the insertion point.
Click **OK**.

SPACEOBJECTS:STYLE
SPACEOBJECTS:BASEAREA

The attribute is placed.

34.

SPACEOBJECTS:STYLE
SPACEOBJECTS:BASEAREA
10' – 0" X 10' – 0"

Place some text below the existing attributes to help you locate the length and width attributes.

You can then modify the text to only show the X.

35.

Define
Attributes

Switch to the Insert ribbon.

Click **Define Attributes**.

36.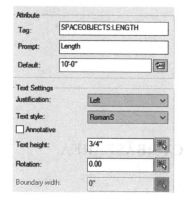

In the Tag field:
Type **SPACEOBJECTS:LENGTH**.
In the Prompt field:
Type **Length**.
In the Default field:
Type **10'- 0"**.
Set the Justification to **Left**.
Set the Text style to **RomanS**.
Disable **Annotative**.
Set the Text height to **3/4"**

37.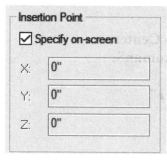

Enable **Specify on-screen** for the insertion point.
Click **OK**.

38.   SPACEOBJECTS:STYLE
SPACEOBJECTS:BASEAREA
1SPACEOBJECTX:LENGTH — 0"

Place below and slightly left of the previous attributes.

Use the text to help you locate the attribute.

39.   SPACEOBJECTS:STYLE
SPACEOBJECTS:BASEAREA
1SPACEOBJCTX:LENGSPACEOBJECTS:LENGTH

Copy the length attribute and place to the right.

40.
Tag:	SPACEOBJECTS:WIDTH
Prompt:	Width
Default:	10'-0"

OK   Cancel

Double click on the copied attribute.
Change the tag to **SPACEOBJECTS:WIDTH**.
Change the prompt to **Width**.
Click **OK**.
Click **ENTER** to exit the command.

41.   SPACEOBJECTS:STYLE
SPACEOBJECTS:BASEAREA
SPACEOBJECTS:LENBASEOBJECTS:WIDTH

Double click on the text and edit to leave only the X.

42.
Door Tag ▾	Door Schedule ▾	Ti
Window Tag ▾	Window Schedule	Se
Room Tag ▾	Room Schedule ▾	El

Call

Wall Tag (Leader)	Wall Schedule
Create Tag	Schedule Styles
Renumber Property Sets	Evaluate Space

Scheduling

Activate the Annotate ribbon.

Select the **Create Tag** tool from the Scheduling panel in the drop-down area.

43.   SPACEOBJECTS:STYLE
SPACEOBJECTS:AREA
SPACEOBJECTS:LENGTH X SPACEOBJECTS:WIDTH

Select the attributes and the text.

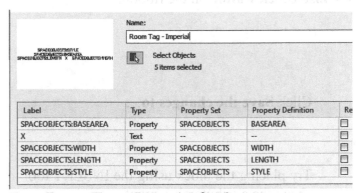

Name:

Room Tag - Imperial

Select Objects
5 items selected

Label	Type	Property Set	Property Definition	Re
SPACEOBJECTS:BASEAREA	Property	SPACEOBJECTS	BASEAREA	☐
X	Text	--	--	☐
SPACEOBJECTS:WIDTH	Property	SPACEOBJECTS	WIDTH	☐
SPACEOBJECTS:LENGTH	Property	SPACEOBJECTS	LENGTH	☐
SPACEOBJECTS:STYLE	Property	SPACEOBJECTS	STYLE	☐

44.   Type **Room Tag – Imperial** for the name.

Verify that the property sets and definitions have been assigned correctly.

If the property definitions are not correct, you may need to double check your attribute definitions.

45. Click **OK**.
Select the center of the attribute list to set the insertion point for the tag.

46. 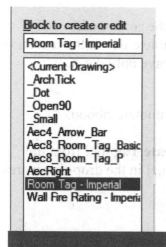 The attributes will be converted into a block.

If the position of the attributes need to be adjusted, insert the Room Tag – Imperial block into the drawing.

47. To change the position of the text or attributes:
Type **BEDIT**.
Save the changes.

48. Select the block in the list.

Click **OK**.

49. The Block Editor will open.
Adjust the position of the text.

50.  Click **Close Block Editor** on the ribbon.

51. → Save the changes to Room Tag - Imperial

→ Discard the changes and close the Block Editor

Click **Save the changes to…**

52.  Try placing the block using the INSERT tool to check the attribute placements.

If you need to adjust the attribute positions, use the BEDIT command again.

53.　　Living Room　　Save as *ex9-23.dwg*.
　　　　00.0 SF
　　10'−0" X 10'−0"

Exercise 9-24:
Create a Room Tag Tool

Drawing Name:　　　palette tools.dwg
Estimated Time:　　5 minutes

This exercise reinforces the following skills:

- Spaces
- Property Set Definitions
- Style Manager
- Property Set Formats
- Attributes
- Add/Delete Scales
- Tags
- Palettes
- Blocks

1.　　Open *palette tools.dwg*.

2.　　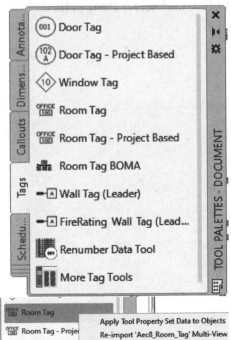　　Change to the **Document** tool palette.

3.　　Locate the **Room Tag** tool on the Document tool palette.

Right click and select **Copy**.

4.
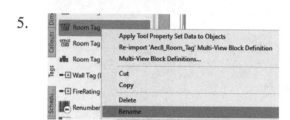
Move the cursor to a blank space on the palette.

Right click and select **Paste**.

5.
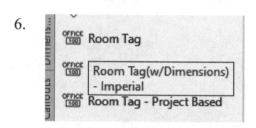
Highlight the copied Room Tag tool.

Right click and select **Rename**.

6.
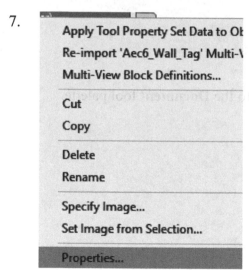
Rename it **Room Tag (w/Dimensions) - Imperial** and move it below the original wall tag.

7.
Highlight the **Room Tag (w/Dimensions) - Imperial.**

Right click and select **Properties**.

Property def location	C:\ProgramData\Autodesk\ACA Tuning beta\enu\Styles\Imperial\Schedule Tables (Imperial).dwg
Tag	
Layer key	SPACENO
Layer key overrides	--
Tag name	Room Tag – Imperial
Tag location	C:\ACA 2023 Fundamentals\ACA 2023 Fundamentals exercises\Lesson_09\schedule tag.dwg
Constrain to object	Yes
Rotate to object	No

8.   Set the Tag Location as *schedule_tag.dwg*.
     Set the Tag Name to **Room Tag - Imperial**.

You may want to create a drawing to store all your custom schedule tags and then use that drawing when setting the tag location.

Click **OK**.

9.   Save as *ex9-24.dwg*.

Schedule Tags

Schedule tags are symbols that connect building objects with lines of data in a schedule table. Schedule tags display data about the objects to which they are attached. There are schedule tag tools with predefined properties provided with the software on the Scheduling tool palette in the Document tool palette set and in the Documentation Tool Catalog in the Content Browser. You can import schedule tags from the Content Browser to add them to your palette.

You can also create custom schedule tags to display specific property set data for objects in the building. After you create a schedule tag, you can drag it to any tool palette to create a schedule tag tool.

Exercise 9-25:
Adding Room Tags

Drawing Name: room_tags.dwg
Estimated Time: 20 minutes

This exercise reinforces the following skills:

- ❏ Tags
- ❏ Tool Palette
- ❏ Content Browser

1.   Open *room_tags.dwg*.

2.   Activate the **A102 03 Room Plan** layout.
       Double click inside the viewport to activate model space.

3.   Select the **Room Tag (w/Dimensions) – Imperial** on the
       tool palette.

4.   Select the family room space.
       Click **ENTER** to place the tag centered in the room.

5.   A small dialog will appear showing all the data attached
       to the space.

       You can modify any field that is not grayed out.

       Click **OK**.

6.

       Zoom in to see the tag.

       Right click and select **Multiple** and then window
       around the entire building.

       Click ENTER to complete the selection.

       This will add space tags to every space.

7.

2 objects were already tagged with the same tag.
Do you want to tag them again?

You will be asked if you want to re-tag spaces which have already been tagged.
Select **No**.

8.

PROPERTY SETS FROM STYLE	–
ManufacturerStyles	
Data source	C:\Users\Elise\SkyDr...
Color	--
Cost	£0.00
Manufacturer	--
Model	--
SpaceStyles	
Data source	C:\Users\Elise\SkyDr...
Style	Den
Length	22'-1"
Width	15'-5"

The small dialog will appear again.

Click **OK**.

All the spaces are now tagged.

Click ENTER to exit the command.

9.

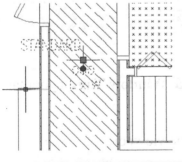

Select the Hallway tag.

Right click and select Properties.

10.

Location on entity

Anchor

On the Properties palette,
select **Anchor**.

11.

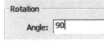

Rotation

Angle: 90

Set the Rotation angle to **90**.
Click **OK**.

12.   Click ESC or left click outside the building model to release the selected tag.

13.  The tag is rotated 90 degrees.

14. Save as *ex9-25.dwg*.

Exercise 9-26:

Adding Door Tags

Drawing Name: door_tags.dwg
Estimated Time: 15 minutes

This exercise reinforces the following skills:

   ❑   Tags

1. Open *door_tags.dwg*.

2. Activate the **Door & Window Tags** layout tab.

3. Activate model space by double-clicking inside the viewport.

4. Activate the Annotate ribbon.
   Select the **Door Tag** tool.

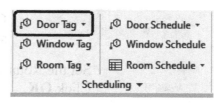

5. 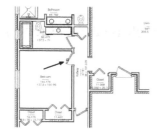 Select a door.

   Click ENTER to center the tag on the door.

6. 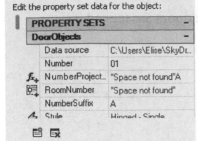 Click **OK**.

Some students may be concerned because they see the Error getting value next to the Room Number field.

If you tag the spaces with a room tag assigning a number, you will not see that error.

7. Right click and select **Multiple**.
Window around the entire building.

All the doors will be selected.

Click **ENTER**.

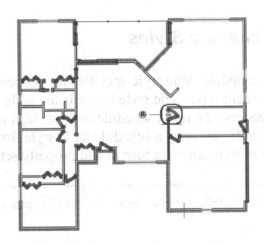

8.
1 object was already tagged with the same tag.
Do you want to tag it again?

Click **No**.

9. Edit the property set data for the objects:

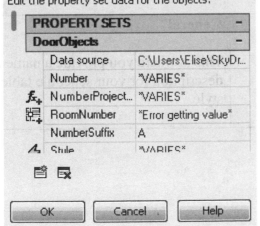

Click **OK.**

10.

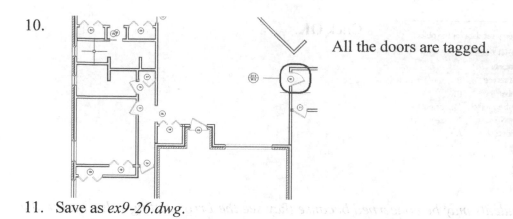

All the doors are tagged.

11. Save as *ex9-26.dwg*.

Extra: *Tag all the windows in the building.*

Schedule Styles

A schedule table style specifies the properties that can be included in a table for a specific element type. The style controls the table formatting, such as text styles, columns, and headers. To use a schedule table style, it must be defined inside the building drawing file. If you copy a schedule table style from one drawing to another using the Style Manager, any data format and property set definitions are also copied.

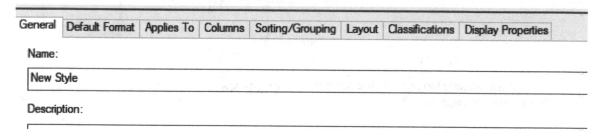

When you create a new Schedule Table Style, there are eight tabs that control the table definition.

Name: New Style Description:	**General** This is where you provide a name and description for your schedule table style.

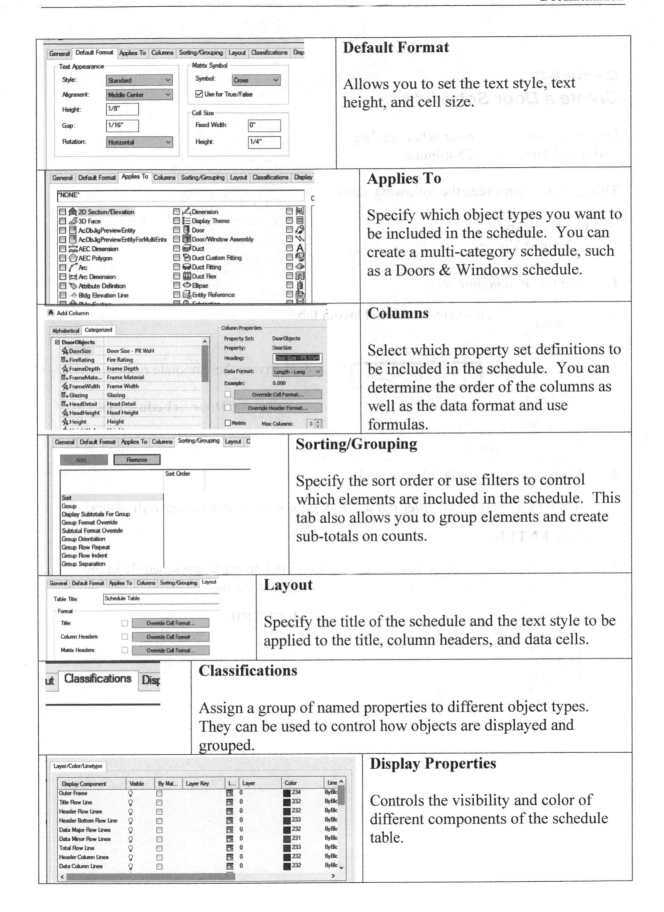

Default Format

Allows you to set the text style, text height, and cell size.

Applies To

Specify which object types you want to be included in the schedule. You can create a multi-category schedule, such as a Doors & Windows schedule.

Columns

Select which property set definitions to be included in the schedule. You can determine the order of the columns as well as the data format and use formulas.

Sorting/Grouping

Specify the sort order or use filters to control which elements are included in the schedule. This tab also allows you to group elements and create sub-totals on counts.

Layout

Specify the title of the schedule and the text style to be applied to the title, column headers, and data cells.

Classifications

Assign a group of named properties to different object types. They can be used to control how objects are displayed and grouped.

Display Properties

Controls the visibility and color of different components of the schedule table.

Exercise 9-27:
Create a Door Schedule

Drawing Name: door schedule.dwg
Estimated Time: 25 minutes

This exercise reinforces the following skills:

    ❑ Schedules
    ❑ Named Views

1.   Open *door schedule.dwg.*

2.   **Model**   Activate the **Model** layout tab.

3.   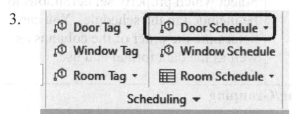   Activate the Annotate ribbon.

    Select the **Door Schedule** tool.

4.   Window around the building model.

    ACA will automatically filter out any non-door entities and collect only doors.

    Click **ENTER**.

5.   Left click to place the schedule in the drawing window.

    Click **ENTER**.

6. *A ? will appear for any doors that are missing tags. So, if you see a ? in the list go back to the model and see which doors are missing tags.*

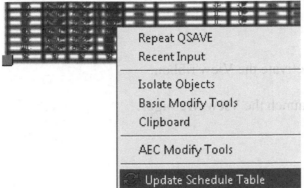

Once you add door tags, you need to update the schedule. Select the schedule, right click and select Update Schedule Table.

7. **Schedules**   Activate the **Schedules** layout tab.

    Double click inside the viewport to activate model space.

8. Rectangular ▾   Activate the Layout ribbon.
    Select the **Rectangular** viewport tool.

9. 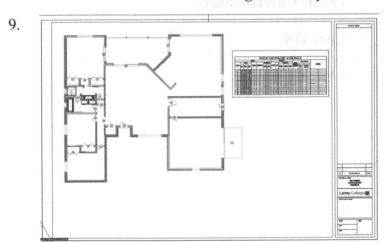   Draw a rectangular viewport next to the floor plan viewport.

    Position the door schedule in the viewport.

10.

Viewport	▾
Layer	Viewport
Display locked	Yes
Annotation scale	1/8" = 1'-0"
Standard scale	Custom
Custom scale	0"
Visual style	2D Wireframe
Shade plot	Hidden

Select the viewport.

Verify that it is on the Viewport layer.

Set the Standard scale to **Custom.**

Set the Display locked to **Yes**.

11. Switch to the MODEL layout.

12. Zoom into the Door Schedule.

13.

Activate the View ribbon.

Launch the **View Manager.**

14. Select **New**.

15.

Type **Door Schedule** for the name.

Enable **Current display**.

Click **OK**.

Close the dialog.

16. Save as *ex9-27.dwg*.

Exercise 9-28:

Create a Door Schedule Style

Drawing Name: schedule_styles.dwg
Estimated Time: 45 minutes

This exercise reinforces the following skills:

- Schedules
- Style Manager
- Text Styles
- Update Schedule

1.   Open *schedule_styles.dwg.*

2.   Open the **MODEL** tab.

3.   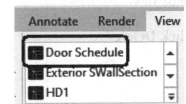   Activate the View ribbon.

Restore the **Door Schedule** view.

4.      Select the door schedule.

Right click and select **Copy Schedule Table Style and Assign…**.

This will create a new style that you can modify.

5.   Select the **General** tab.

Rename **Custom Door Schedule**.

6.   Select the **Columns** tab.

7. 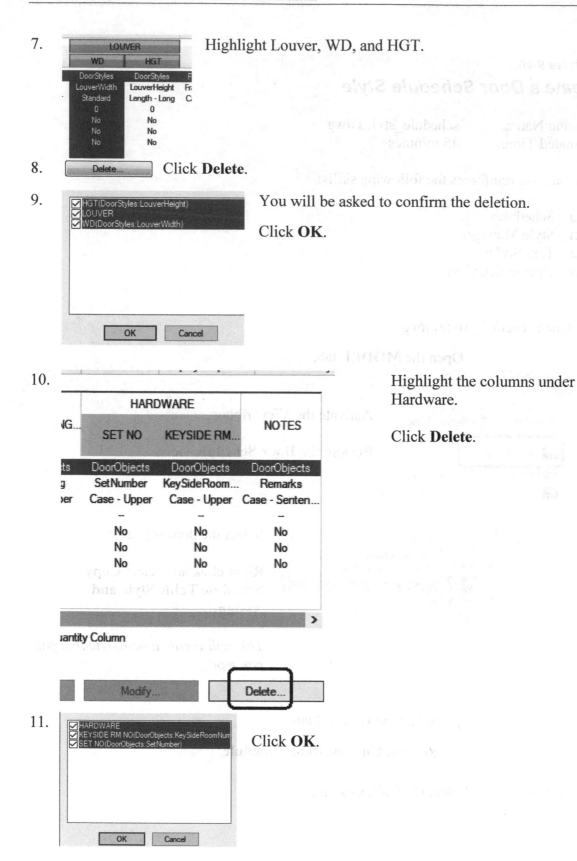 Highlight Louver, WD, and HGT.

8. Click **Delete**.

9. You will be asked to confirm the deletion.

   Click **OK**.

10. Highlight the columns under Hardware.

    Click **Delete**.

11. Click **OK**.

12.
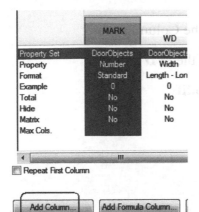

Highlight **MARK**.

Click the **Add Column** button.

13.
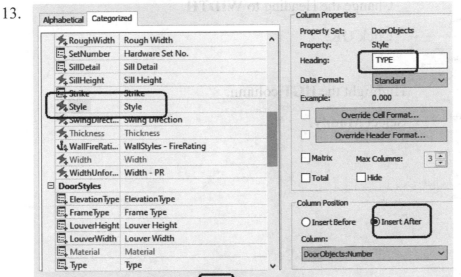

Scroll down until you see Style.

Highlight **Style**.

Confirm that the Column Position will be inserted after the DoorObjects: Number. Change the Heading to **TYPE**.

Click **OK**.

14.
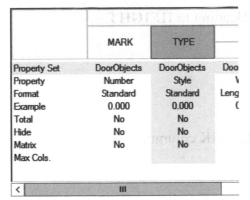

The new column is now inserted.

15.

Select the Layout tab.

Change the Table Title to **DOOR SCHEDULE**.

16.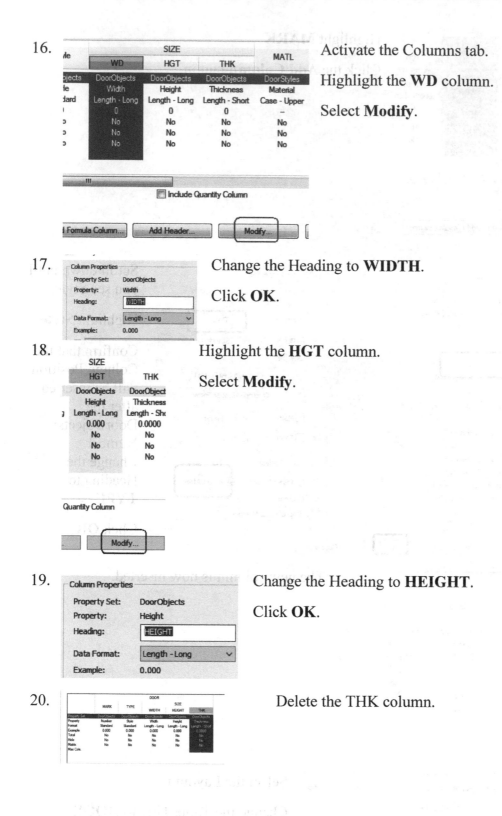

Activate the Columns tab.

Highlight the **WD** column.

Select **Modify**.

17.

Change the Heading to **WIDTH**.

Click **OK**.

18.

Highlight the **HGT** column.

Select **Modify**.

19.

Change the Heading to **HEIGHT**.

Click **OK**.

20.

Delete the THK column.

21.
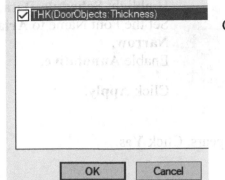
Click **OK**.

22. Click **OK** to close the dialog.

23.

DOOR			
MARK	TYPE	SIZE	
		WIDTH	HEIGHT
1	Hinged – Single	2'-6"	6'-8"
2	Cased Opening – Halfround	3'-0"	6'-8"
3	Cased Opening – Halfround	2'-6"	6'-8"
4	Pocket – Single	2'-4"	6'-8"
5	Pocket – Single	2'-4"	6'-8"
6	Sliding – Double – Full Lite	10'-0"	6'-8"
7	Sliding – Double – Full Lite	5'-4"	6'-8"
8	Sliding – Double – Full Lite	5'-4"	6'-8"
9	Overhead – Sectional	16'-0"	6'-8"
10	Hinged – Single – Exterior	2'-6"	6'-8"
11	Hinged – Single	2'-6"	6'-8"
12	Hinged – Single	2'-6"	6'-8"
13	Hinged – Single	2'-6"	6'-8"
14	Hinged – Single	2'-6"	6'-8"
15	Hinged – Single – Exterior	3'-0"	6'-8"
16	Bifold – Double	4'-0"	6'-8"
17	Bifold – Single	3'-0"	7'-0"
18	Bifold – Double	4'-6"	6'-8"
19	Bifold – Double	4'-6"	6'-8"
20	Bifold – Double	4'-6"	6'-8"
21	Bifold – Double	4'-6"	6'-8"

The schedule updates.

If there are question marks in the schedule, that indicates doors which remain untagged.

24.

Select the schedule.
Select **Update** from the ribbon.
Click **ESC** to release the selection.

25. Activate the Home ribbon.

Select the **Text Style** tool on the Annotation panel.

26. RomanS
Schedule-Data
Schedule-Header
Schedule-Title

Notice there are text styles set up for schedules.
Highlight each one to see their settings.

27. 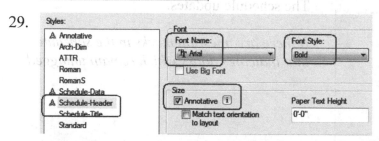 Highlight **Schedule-Data.**
Set the Font Name to **Arial Narrow.**
Enable **Annotative.**

Click **Apply.**

28. If this dialog appears, Click **Yes.**

29. Highlight Schedule-Header.
Set the Font Name to **Arial.**
Set the Font Style to **Bold.**
Enable **Annotative.**

Click **Apply.**

30. 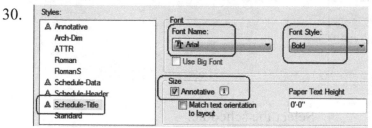 Highlight Schedule-Title.
Set the Font Name to **Arial.**
Set the Font Style to **Bold.**
Enable **Annotative.**

Click **Apply** and **Close.**

31. Note that the schedule automatically updates.

DOOR SCHEDULE

| MARK | TYPE | SIZE | |
		WIDTH	HEIGHT
	DOOR		
1	Hinged - Single	2'-8"	6'-8"
2	Cased Opening - Halfround	3'-0"	6'-8"
3	Cased Opening - Halfround	2'-8"	6'-8"
4	Pocket - Single	2'-4"	6'-8"
5	Pocket - Single	2'-4"	6'-8"
6	Sliding - Double - Full Lite	10'-0"	6'-8"
7	Sliding - Double - Full Lite	6'-4"	6'-8"
8	Sliding - Double - Full Lite	6'-4"	6'-8"
9	Overhead - Sectional	16'-0"	6'-8"
10	Hinged - Single - Exterior	2'-8"	6'-8"
11	Hinged - Single	2'-8"	6'-8"
12	Hinged - Single	2'-8"	6'-8"
13	Hinged - Single	2'-8"	6'-8"
14	Hinged - Single	2'-0"	6'-8"
15	Hinged - Single - Exterior	3'-0"	6'-8"
16	Bifold - Double	4'-0"	6'-8"
17	Bifold - Single	3'-0"	7'-0"
18	Bifold - Double	4'-6"	6'-8"
19	Bifold - Double	4'-6"	6'-8"
20	Bifold - Double	4'-6"	6'-8"
21	Bifold - Double	4'-6"	6'-8"

32. Save as *ex9-28.dwg.*

Exercise 9-29:

Create a Room Schedule

Drawing Name: room_schedule.dwg
Estimated Time: 30 minutes

This exercise reinforces the following skills:

- ❑ Schedules
- ❑ Tool Palettes
- ❑ Creating Tools
- ❑ Schedule Styles
- ❑ Property Set Definitions
- ❑ Add scale to annotative objects
- ❑ Add viewport

1. Open *room_schedule.dwg*.

2.  Activate the Manage ribbon.

   Select **Style Manager**.

3. Locate the Schedule Table Styles category under Documentation.

   - Architectural Objects
   - Documentation Objects
     - 2D Section/Elevation Styles
     - AEC Dimension Styles
     - Calculation Modifier Styles
     - Display Theme Styles
     - Property Data Formats
     - Property Set Definitions
     - Schedule Table Styles
       - Custom Door Schedule
       - Door Schedule
       - Room Finish Schedule
       - Standard

4. Highlight the Schedule Table Styles category.

   Right click and select **New**.

5. On the General tab:

General	Default Format	Applies To

   Name:
   Room Schedule

   Description:
   Uses Space Information

   Type **Room Schedule** for the Name.

   Type **Uses Space Information** for the Description.

6.

Select the Applies To tab.

Enable **Space**.

Note that Spaces have Classifications available which can be used for sorting.

7.

Select the Columns tab.

Select **Add Column**.
Select **Style**.

8.

Change the Heading to **Room Name**.

Set the Data Format to **Case – Sentence**.

Click **OK**.

If your property set definitions have not been modified, you will not be able to access these properties.

9.

Select **Add Column**.

10.

Select the Alphabetical tab to make it easier to locate the property definition.

Locate and select **Length**.

Click **OK**.

11.

Select **Add Column**.

12.

Select the Alphabetical tab to make it easier to locate the property definition.

Locate and select **Width**.

Click **OK**.

13.

Select Add Column.

14. 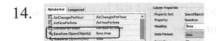 Select the Alphabetical tab to make it easier to locate the property definition.

Select **BaseArea**.

Change the Heading to **Area**.

Click **OK**.

15. The columns should appear as shown.

	Room Name	Length	Width	Area
Property Set	SpaceObjects	SpaceObjects	SpaceObjects	SpaceObjects
Property	Style	Length	Width	BaseArea
Format	Case - Senten...	Length - Short	Length - Short	Area
Example				SF
Total	No	No	No	No
Hide	No	No	No	No
Matrix	No	No	No	No
Max Cols.				

16. Select the Layout tab.

Change the Table Title to **Room Schedule.**

Table Title: Room Schedule

Format

Title: ☐ Override Cell Format...

Column Headers: ☐ Override Cell Format...

Matrix Headers: ☐ Override Cell Format...

17. 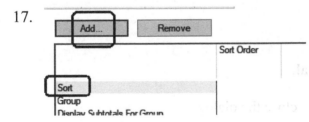 Select the **Sorting/Grouping** tab. Highlight **Sort**. Select the **Add** button.

18. 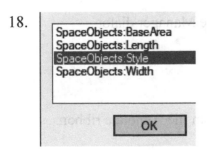 Select **SpaceObjects: Style** to sort by Style name. Click **OK**.

SpaceObjects:BaseArea
SpaceObjects:Length
SpaceObjects:Style
SpaceObjects:Width

OK

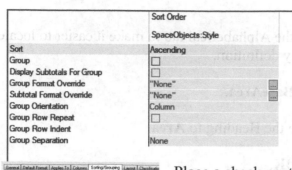

The panel should fill in with the sort information.

19. Place a check next to **Group**.
Place a check next to **Display Subtotals for Group.**

Note: To enable subtotals modify at least one column on the columns tab to include a total.

A small red note will appear at the bottom of the dialog.

20. Select the Columns tab.

Modify...

Highlight **Area** and select **Modify**.

21. Enable **Total.**

Click **OK** to close the dialog.

Click **OK** to close the Style Manager dialog.

22. Launch the Annotation Tools palette from the Annotate ribbon.

23.  Select the Schedules tab.

24. 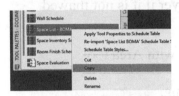 Highlight the Space List – BOMA.

Right click and select Copy.

25. 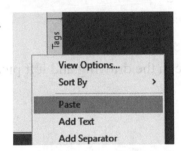 Move the mouse off the tool.

Right click and select **Paste**.

26. 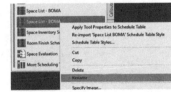 Select the copied tool.

Right click and select **Rename**.

27. 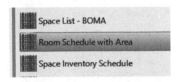 Type **Room Schedule with Area** for the name.

28.  Highlight the new tool.

Right click and select **Properties**.

29.

In the Style field, select the **Room Schedule** style which you just created.

Click **OK**.

30.

Activate the Model tab.

Thaw the **G-Area-Spce** layer if it is not thawed.

31.

Room Name	Length	Width	Area
Bathroom	5'-3 3/4"	15'-0 1/2"	71.4 SF
	5'-11 3/4"	11'-6 1/4"	61.3 SF
			132.7 SF
Bedroom	18'-10 3/4"	15'-0"	253.4 SF
	13'-4 1/4"	15'-0"	189.8 SF
	12'-7 1/2"	11'-6 1/4"	145.4 SF
			568.6 SF
Closet	5'-5 3/4"	2'-6 3/4"	14.0 SF
	3'-8 1/4"	3'-1"	11.4 SF
	2'-5 1/2"	5'-2"	12.7 SF
	5'-10 1/2"	2'-5"	14.2 SF
	5'-0 3/4"	2'-6"	12.7 SF
	1'-4 1/4"	3'-0"	4.0 SF
			69.1 SF
Den	13'-7 1/2"	20'-0 3/4"	252.4 SF
			252.4 SF
Dining room	10'-3 1/4"	20'-2"	181.7 SF
			181.7 SF
Garage	19'-11 1/2"	20'-8"	412.5 SF
			412.5 SF
Hallway	15'-11"	3'-5 1/4"	54.7 SF
			54.7 SF
Kitchen	9'-2"	9'-6 3/4"	85.5 SF
			85.5 SF
Laundry room	20'-9 1/4"	7'-8 1/4"	159.7 SF
			159.7 SF
Livingroom	27'-6 3/4"	25'-6 1/4"	615.6 SF
			615.6 SF
			2532.5 SF

Select the Room Schedule with Area tool.

Window around the building.

Click ENTER.

Pan over to an empty space in the drawing and left pick to place the schedule.

Click ENTER.

Room Schedule			
Room Name	Length	Width	Area
?	?	?	? SF
	?	?	? SF
	?	?	? SF
Bathroom	11'–5 1/2"	5'–11"	60.5 SF
	15'–0"	5'–2 1/2"	69.2 SF
			129.7 SF
Bedroom	15'–0"	16'–4 3/4"	245.9 SF
	11'–5 1/2"	12'–7"	144.2 SF
	15'–0"	13'–3 3/4"	169.4 SF
			559.5 SF
Closet	5'–1 3/4"	1'–11 3/4"	10.2 SF
	5'–5 1/4"	1'–11 1/2"	10.7 SF
	5'–0 1/4"	1'–11 1/2"	9.9 SF
	3'–9"	2'–11 1/4"	11.1 SF
	5'–9 1/4"	1'–11 3/4"	11.4 SF
			53.2 SF
Den	20'–0"	13'–7"	250.1 SF
			250.1 SF
Dining room	21'–2 1/2"	10'–3 3/4"	178.7 SF
			178.7 SF
Garage	20'–8"	19'–11 1/2"	412.5 SF
			412.5 SF
Hallway	3'–0"	15'–11 1/4"	47.8 SF
			47.8 SF
Kitchen	13'–0 3/4"	13'–1"	85.4 SF
			85.4 SF
Laundry room	20'–8"	7'–7"	156.7 SF
			156.7 SF
Livingroom	25'–5 1/2"	27'–6 1/2"	597.5 SF
			597.5 SF
			2471.2 SF

If there are any ? symbols in the table, you have some spaces which are unassigned.

You may have placed duplicate spaces or empty spaces.

It may be easier to select all the spaces using QSELECT and delete them and then replace all the spaces.

32.

Activate the **A102 03 ROOM PLAN** layout tab.

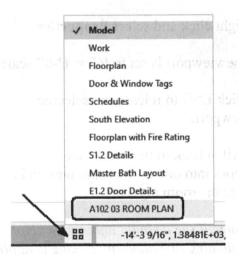

33.

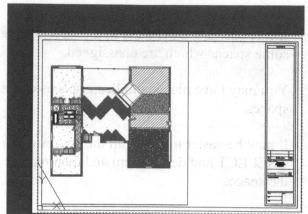

Select the viewport.

Adjust the size and position of the viewport, so that there is an empty area on the right for another viewport.

34.

Activate the viewport and freeze the G-Area-Spce layer in the viewport.

If you don't see the space tags, it is because you need to add the correct annotative scale for the tags.

Double click outside the viewport to deactivate.
Select the viewport for the floor plan.

35.

Clipped	No
Display locked	Yes
Annotation scale	1/4" = 1'-0"
Standard scale	1/4" = 1'-0"
Custom scale	1/32"

Right click and select **Properties**.

The viewport is set to ¼" = 1'-0" scale.

Click **ESC** to release the selected viewport.

36.

Switch back to the Model tab.
Zoom into one of the room tags and hover over the room tag.

You will see a single annotative glyph indicating one annotative scale is applied to the tag.

37.

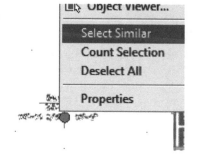

Switch to the Annotate ribbon.
Select a room tag.

Right click and select **Select Similar**.

This will select all the room tags in the window.

38.

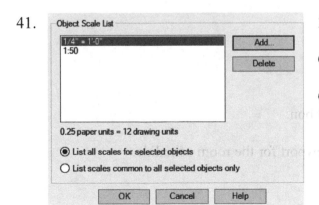

On the ribbon:
Click **Add/Delete Scales**.

39.

Click **Add**.

Object Scale List

1:50

Add...

Delete

1 paper unit = 50 drawing units

⦿ List all scales for selected objects
◯ List scales common to all selected objects only

40.

Scale List

1:25
1/64" = 1'-0"
1/32" = 1'-0"
1/16" = 1'-0"
3/32" = 1'-0"
1/8" = 1'-0"
3/16" = 1'-0"
1/4" = 1'-0"
3/8" = 1'-0"
1/2" = 1'-0"
3/4" = 1'-0"
1" = 1'-0"
1 1/2" = 1'-0"
3" = 1'-0"

OK   Cancel

Highlight the ¼" = 1"-0" scale.

Click **OK**.

41.

Object Scale List

1/4" = 1'-0"
1:50

Add...

Delete

0.25 paper units = 12 drawing units

⦿ List all scales for selected objects
◯ List scales common to all selected objects only

OK   Cancel   Help

The scale is added to the list.

Click **OK**.

Click **ESC** to release the selection.

42.

Now if you hover over the tag, you will see two annotative glyphs, indicating more than one scale is enabled.

43.

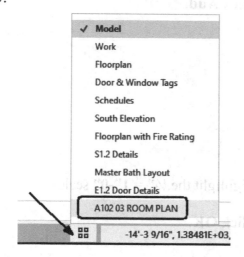

Activate the **A102 03 ROOM PLAN** layout tab.

All the room tags are now visible.

44.

Activate the Layout ribbon.

Place a rectangular viewport for the room schedule.

Rectangular

45.

Use Zoom Extents and Zoom Window to center the room schedule in the viewport.

46.

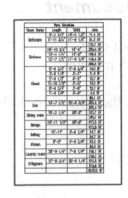

Save as *ex9-29.dwg*.

Extra: Create a named view for the room schedule.

Create a different room schedule style.

Create a new layout with the floor plan and the new room schedule.

Sheet Set Manager

The Sheet Set Manager provides a list of layout tabs available in a drawing or set of drawings and allows you to organize your layouts to be printed.

The Sheet Set Manager allows you to print to a printer as well as to PDF. Autodesk does not license their PDF creation module from Adobe. Instead, they have a partnership agreement with a third-party called Bluebeam. This means that some fonts may not appear correct in your PDF print.

Exercise 9-30:

Creating a PDF Document

Drawing Name: pdf.dwg
Estimated Time: 15 minutes

This exercise reinforces the following skills:

- ❏   Printing to PDF
- ❏   Sheet Set Manager
- ❏   Xref Manager

1.   Open *pdf.dwg*.

2.   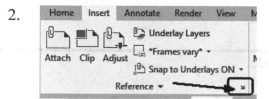   Switch to the **Insert** ribbon.
Launch the XREF Manager.

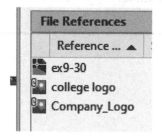   The file has a couple of image files used in the titleblock.

We need to re-path the image files since you downloaded the files and need to point the reference to the correct file locations.

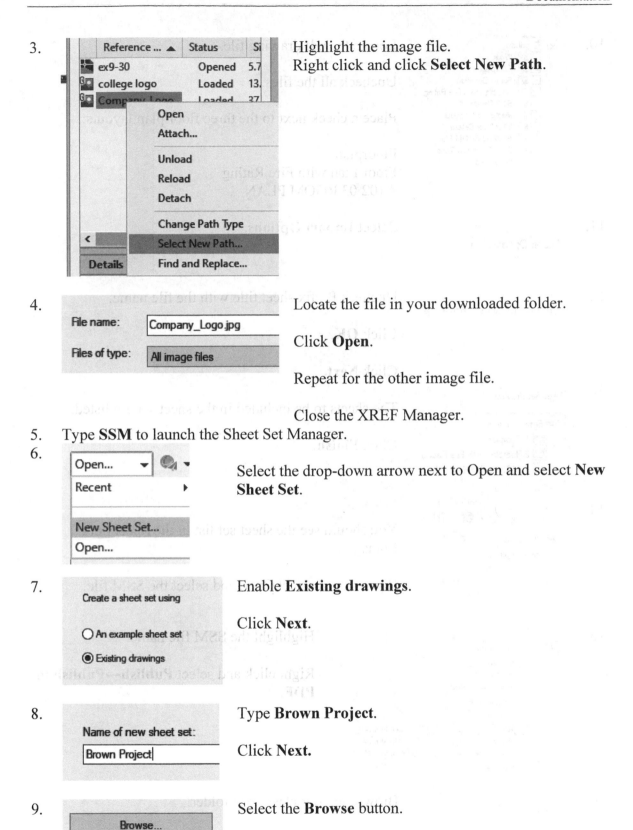

3. Highlight the image file.
    Right click and click **Select New Path**.

4. Locate the file in your downloaded folder.

    Click **Open**.

    Repeat for the other image file.

    Close the XREF Manager.

5. Type **SSM** to launch the Sheet Set Manager.

6. Select the drop-down arrow next to Open and select **New Sheet Set**.

7. Enable **Existing drawings**.

    Click **Next**.

8. Type **Brown Project**.

    Click **Next**.

9. Select the **Browse** button.

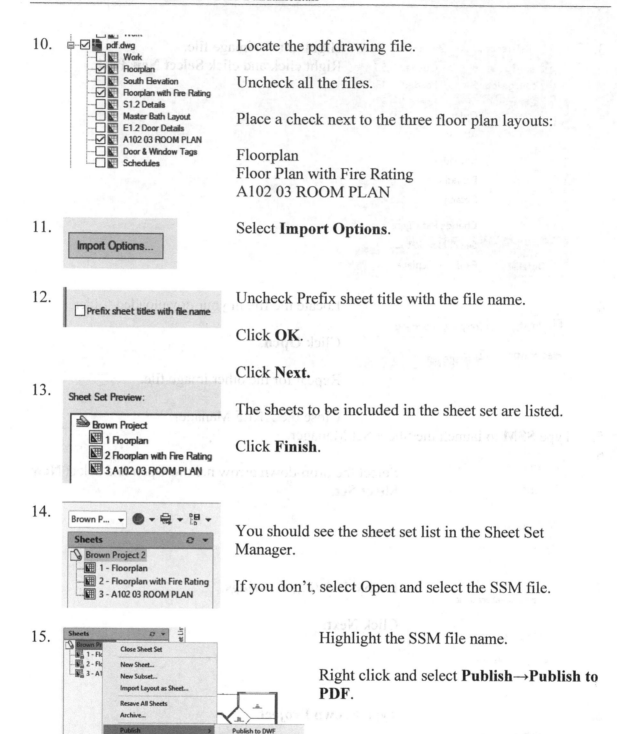

10. Locate the pdf drawing file.

   Uncheck all the files.

   Place a check next to the three floor plan layouts:

   Floorplan
   Floor Plan with Fire Rating
   A102 03 ROOM PLAN

11. Select **Import Options**.

12. Uncheck Prefix sheet title with the file name.

   Click **OK**.

   Click **Next.**

13. The sheets to be included in the sheet set are listed.

   Click **Finish**.

14. You should see the sheet set list in the Sheet Set Manager.

   If you don't, select Open and select the SSM file.

15. Highlight the SSM file name.

   Right click and select **Publish→Publish to PDF**.

16.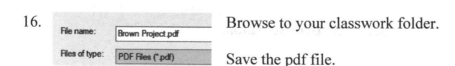

   Browse to your classwork folder.

   Save the pdf file.

17. 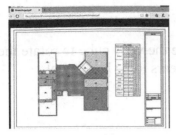 Browse to where the file was saved using Windows File Explorer.

Open the PDF file and review.

Save as *ex9-30.dwg*.

Legends

Component legends are used to display symbols that represent model components, such as doors and windows. Some architectural firms may already have tables with these symbols pre-defined. In which case, you can simply add them to your drawing as an external reference or as a block. However, if you want to generate your own legends using the components in your ACA model, you do have the tools to do that as well.

Exercise 9-31:

Create Door Symbols for a Legend

Drawing Name: legend.dwg
Estimated Time: 40 minutes

This exercise reinforces the following skills:

- ❑ Doors
- ❑ Images
- ❑ Object Viewers

1. Open *legend.dwg*.

2. 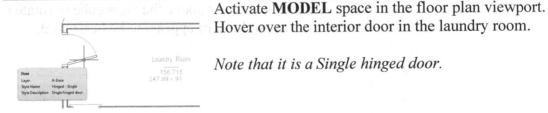 Select the **A102 03 Floor Plan** layout.

Activate **MODEL** space in the floor plan viewport.

3. Hover over the interior door in the laundry room.

Note that it is a Single hinged door.

4. 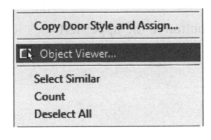 Select the interior door in the laundry room.

Right click and select **Object Viewer**.

5.

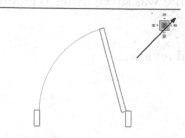

Use the arrows above the view cube to rotate the view so the door appears to be horizontal.

6.

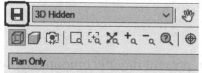

Select the **Save** button in the upper left.

7.

File name: single-hinged

Files of type: PNG (*.png)

Browse to your work folder.

Type **single-hinged** for the file name.

Click **Save**.
Close the Object Viewer.

8.

Hover over the exterior door in the laundry room.

Note that it is a Single exterior hinged door.

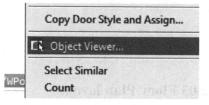

Door
Layer   A-Door
Style Name   Hinged - Single - Exterior
Style Description   Single exterior door

9.

Copy Door Style and Assign...

Object Viewer...

Select Similar

Count

Select the exterior door in the laundry room.

Right click and select **Object Viewer**.

10.

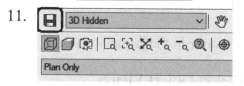

Use the arrows above the view cube to rotate the view so the door appears to be horizontal.

11.

3D Hidden

Plan Only

Select the **Save** button in the upper left.

12.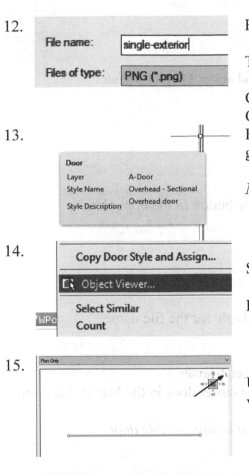

Browse to your work folder.

Type **single-exterior** for the file name.

Click **Save**.
Close the Object Viewer.

13. Hover over the sectional overhead door in the garage.

Note that it is an Overhead-Sectional door.

14. Select the door.

Right click and select **Object Viewer**.

15. Use the arrows above the view cube to rotate the view so the door appears to be horizontal.

16.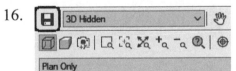

Select the **Save** button in the upper left.

17. Browse to your work folder.

Type **overhead-sectional** for the file name.

Click **Save**.
Close the Object Viewer.

18. Hover over the closet door near the home entrance.

Note that it is a Bifold-single door.

19. 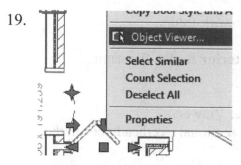 Select the door.

Right click and select **Object Viewer**.

20. 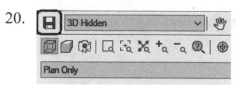 Select the **Save** button in the upper left.

21. 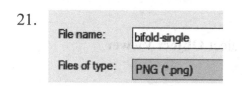 Browse to your work folder.

Type **bifold-single** for the file name.

Click **Save**.
Close the Object Viewer.

22. 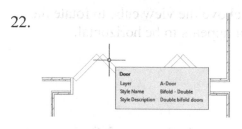 Hover over the closet door in the Master bedroom.

Note that it is a Bifold-Double door.

23. 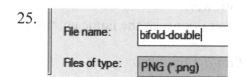 Select the door.

Right click and select **Object Viewer**.

24. Select the **Save** button in the upper left.

25. Browse to your work folder.

Type **bifold-double** for the file name.

Click **Save**.
Close the Object Viewer.

26.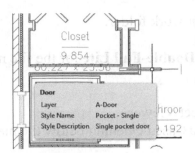

Hover over the pocket door below the master bedroom.

Note that it is a Single Pocket door.

27. Select the door.

Right click and select **Object Viewer**.

28. Select the **Save** button in the upper left.

29.

| File name: | single-pocket |
| Files of type: | PNG (*.png) |

Browse to your work folder.

Type **single-pocket** for the file name.

Click **Save**.
Close the Object Viewer.

30. Hover over the sliding door in the master bedroom.

Note that it is a Sliding-Double-Full Lite door.

31. Select the door.

Right click and select **Object Viewer**.

32.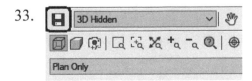

Use the arrows above the view cube to rotate the view so the door appears to be horizontal.

33. Select the **Save** button in the upper left.

34.

File name:	Sliding-Double-Full Lite
Files of type:	PNG (*.png)

Browse to your work folder.

Type **Sliding-Double-Full Lite** for the file name.

Click **Save**.
Close the Object Viewer.

35.

Closet

Door	
Layer	A-Wall-Open
Style Name	Cased Opening - Halfround
Style Description	Cased halfround opening

Hover over the Cased Opening below the master bedroom.

Note that it is a Cased halfround Opening.

36.

Object Viewer...

Select Similar
Count Selection
Deselect All

Properties

Select the opening.

Right click and select **Object Viewer**.

37.

3D Hidden

Plan Only

Select the **Save** button in the upper left.

38.

File name:	cased-opening-halfround
Files of type:	PNG (*.png)

Browse to your work folder.

Type **cased-opening-halfround** for the file name.

Click **Save**.
Close the Object Viewer.

39. Save as *ex9-31.dwg*.

Exercise 9-32:
Convert Images to Blocks

Drawing Name: blocks.dwg
Estimated Time: 45 minutes

This exercise reinforces the following skills:

- Images
- Blocks
- Attach

1. Open *blocks.dwg.*

2. Activate the **MODEL** layout tab.

3. 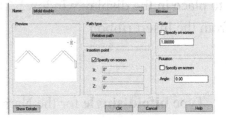 Activate the Insert ribbon.

   Select **Attach**.

4. Locate the **bifold-double** image file.
   Click **Open**.

   File name: bifold-double.png

   Files of type: All image files

5. Enable **Specify Insertion point**.
   Disable **Scale**.
   Disable **Rotation**.

   Click **OK**.

   Left click to place in the display window.

6. 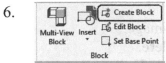 Select **Create Block** from the Insert ribbon.

7.

Type **bifold-double** for the Name.

Select a corner of the image as the base point.
Select the image.
Enable **Delete.**
Enable **Scale Uniformly.**
Disable **Allow Exploding.**
Click **OK.**

8.

Select **Attach**.

9.

Locate the **bifold-single** image file.
Click **Open.**

10.

Enable **Specify Insertion point**.
Disable **Scale**.
Disable **Rotation**.

Click **OK**.

Left click to place in the display window.

11.

Select **Create Block** from the Insert ribbon.

12.

Type **bifold-single** for the Name.

Select a corner of the image as the base point.
Select the image.
Enable **Delete.**
Enable **Scale Uniformly.**
Disable **Allow Exploding**.

Click **OK**.

13.

Select **Attach**.

14.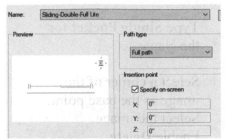

Locate the **Sliding-Double-Full Lite** image file.
Click **Open**.

15.

Enable **Specify Insertion point**.
Disable **Scale**.
Disable **Rotation**.

Click **OK**.

Left click to place in the display window.

16.

Select **Create Block** from the Insert ribbon.

17.

Type **Sliding-Double-Full Lite**
for the Name.

Select a corner of the image as the
base point.
Select the image.
Enable **Delete.**
Enable **Scale Uniformly**.
Disable **Allow Exploding**.

Click **OK**.

18.

Select **Attach**.

19.

Locate the **Single Pocket** image file.
Click **Open**.

File name: single-pocket.png

Files of type: All image files

20.

Enable **Specify Insertion point**.
Disable **Scale**.
Disable **Rotation**.

Click **OK**.

Left click to place in the display
window.

21. 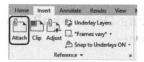 Select **Create Block** from the Insert ribbon.

22. 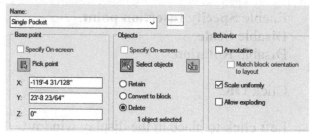 Type **Single Pocket** for the Name.

    Select a corner of the image as the base point.
    Select the image.
    Enable **Delete.**
    Enable **Scale Uniformly**.
    Disable **Allow Exploding**.

    Click **OK**.

23. 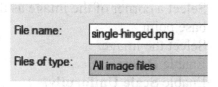 Select **Attach**.

24. Locate the **single-hinged** image file.
    Click **Open**.

    File name: single-hinged.png

    Files of type: All image files

25. 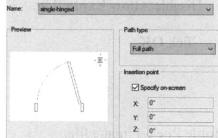 Enable **Specify Insertion point**.
    Disable **Scale**.
    Disable **Rotation**.

    Click **OK**.

    Left click to place in the display window.

26. 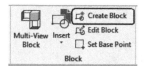 Select **Create Block** from the Insert ribbon.

27.

Type **single-hinged** for the Name.

Select a corner of the image as the base point.
Select the image.
Enable **Delete.**
Enable **Scale Uniformly**.
Disable **Allow Exploding**.

Click **OK**.

28.

Select **Attach**.

29.

File name: single-exterior.png

Files of type: All image files

Locate the **single-exterior** image file.
Click **Open**.

30.

Enable **Specify Insertion point**.
Disable **Scale**.
Disable **Rotation**.

Click **OK**.

Left click to place in the display window.

31.

Select **Create Block** from the Insert ribbon.

32.

Type **single-exterior** for the Name.

Select a corner of the image as the base point.
Select the image.
Enable **Delete.**
Enable **Scale Uniformly**.
Disable **Allow Exploding**.

Click **OK**.

33.

Select **Attach**.

34.

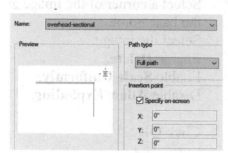

Locate the **overhead-sectional** image file.
Click **Open**.

35.

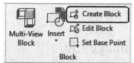

Enable **Specify Insertion point**.
Disable **Scale**.
Disable **Rotation**.

Click **OK**.

Left click to place in the display window.

36.

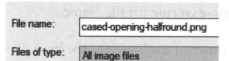

Select **Create Block** from the Insert ribbon.

37.

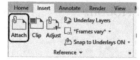

Type **overhead-sectional** for the Name.

Select a corner of the image as the base point.
Select the image.
Enable **Delete.**
Enable **Scale Uniformly**.
Disable **Allow Exploding**.

Click **OK**.

38.

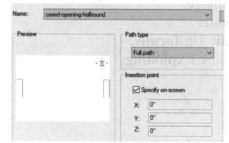

Select **Attach**.

39.

Locate the **cased-opening-halfround** image file.
Click **Open**.

40.

Enable **Specify Insertion point**.
Disable **Scale**.
Disable **Rotation**.

Click **OK**.

Left click to place in the display window.

41.   Select **Create Block** from the Insert ribbon.

42.   Type **cased-opening-halfround** for the Name.

Select a corner of the image as the base point.
Select the image.
Enable **Delete.**
Enable **Scale Uniformly.**
Disable **Allow Exploding**.

Click **OK**.

43. Save as *ex9-32.dwg*.

Exercise 9-33:

Create a Door Legend

Drawing Name: legend2.dwg
Estimated Time: 30 minutes

This exercise reinforces the following skills:

- Legends
- Tables
- Blocks
- Text Styles
- Viewports
- Layouts

1. Open *legend2.dwg*.

2.   Activate the **MODEL** layout tab.

3. 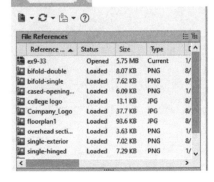 In order for the blocks to appear properly, you need to re-path all the image files.

Bring up the XREF Manager and assign the images to the correct file location.

4.

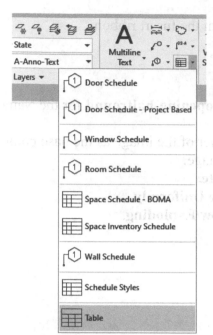

Activate the Home ribbon.

Select the **Table** tool located in the drop-down on the Annotation panel.

5.

Set the Columns to **2**.
Set the Data rows to **9.**

Click **OK**.

6.

DOOR LEGEND

Left click to place the table next to the door schedule.

Type **DOOR LEGEND** as the table header.

7.

Highlight the second row, first cell.
Right click and select **Insert Block**.

8.

Name: bifold-double Browse...
Path:
Properties
Scale: 1.00000
☑ AutoFit
Rotation angle: 0.00
Overall cell alignment: Middle Center
OK Cancel Help

Select **bifold-double** from the drop-down list.
Enable **Auto-fit**.
Set the Overall cell alignment to **Middle Center**.
Click **OK**.

9.

The image appears in the cell.

10.

Bifold Double

Type **Bifold Double** in the second cell, second row.

Change the text height to **1/16"**.

Adjust the width of the columns.

11.

Block Field Formula Manage Cell Contents
Insert

Highlight the third row, first cell.
Select **Insert Block** from the ribbon.

12.

Name: bifold-single Browse...
Path:
Properties
Scale: 1.00000
☑ AutoHt
Rotation angle: 0.00
Overall cell alignment: Middle Center

Select **bifold-single** from the drop-down list.
Enable **Auto-fit**.
Set the Overall cell alignment to **Middle Center**.
Click **OK**.

13.

Bifold Double

The image appears in the cell.

14.

Bifold Double

Bifold Single

Type **Bifold Single** in the second cell, third row.

Change the text height to **1/16"**.

15.

Highlight the fourth row, first cell.
Right click and select **Insert Block**.

16.

Select **single-hinged** from the drop-down list.
Enable **Auto-fit**.
Set the Overall cell alignment to **Middle Center**.
Click **OK**.

17.

Bifold Double

Bifold Single

The image appears in the cell.

18.

Bifold Double

Bifold Single

Single Hinged

Type **Single Hinged** in the second cell, fourth row.

Change the text height to **1/16"**.

19.

DOOR LEGEND

Bifold Double

Bifold Single

Single Hinged

Single Exterior

Single Pocket

Sliding Double Full Lite

Overhead Sectional

Cased Opening – Halfround

Continue inserting the blocks and text until the door legend is completed as shown.

Unsaved Layer State

A-Anno-Schd-Bdr

Place the table on the **A-Anno-Schd-Bdr** layer.

Type IMAGEFRAME and then 0 to turn off the border around the images.

20.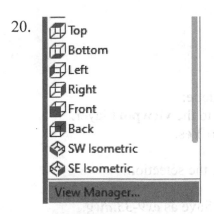

Go to the View ribbon.

Launch the **View Manager**.

21. New... Select **New**.

22.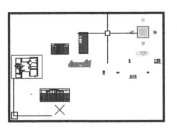

Type **Door Legend** in the View name field.
Enable **Define window**.

Window around the legend table.
Click **ENTER**.
Click **OK**.
Close the View Manager.

23. **E1.2 Door Details**

Activate the **E1.2 Door Details** layout.

24. Rectangular

Activate the Layout ribbon.

Select the **Rectangular** viewport tool.

25.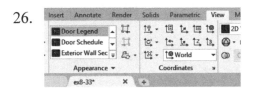

Draw the rectangle to place the viewport on the layout sheet.

Double click inside the viewport to activate MODEL space.

26. Activate the View ribbon.

Double click on the **Door Legend** view.

The display in the viewport updates with the door legend.

27.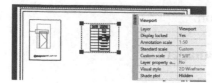

Select the viewport.

On the Properties palette:
Assign the viewport to the **viewport** layer.
Set Display locked to **Yes**.

Click ESC to release the selection.

28.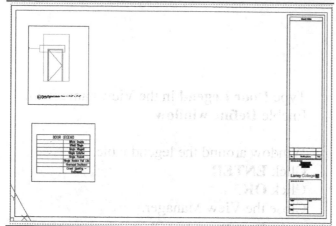

Save as *ex9-33.dwg*.

Extra:

1) Update the view titles with the correct sheet numbers based on the sheets where the views are placed.

2) Create additional elevation and section views for the kitchen, the second bathroom, the exterior elevations, and windows.

3) Add the new views to existing sheets or new sheets.

4) Add window tags to all the windows.

5) Create a window schedule.

6) Create a window legend.

7) Create a detail view for window framing.

8) Create a callout to the window framing detail view.

9) Create a section elevation view of an interior wall.

10) Add keynotes to the interior wall section elevation view.

11) Update the attributes on each title block with the correct sheet number and title information.

Trace

Trace provides a safe space to add changes to a drawing in the web and mobile apps without altering the existing drawing. The analogy is of a virtual collaborative tracing paper that is laid over the drawing that allows collaborators to add feedback on the drawing.

Create traces in the web and mobile apps, then send or share the drawing to collaborators so they can view the trace and its contents.

Traces can be viewed in the desktop application but can only be created or edited in the mobile apps.

Exercise 9-34:
Use Trace

Drawing Name: trace.dwg
Estimated Time: 5 minutes

This exercise reinforces the following skills:

- ❑ markups
- ❑ Trace

You can share a drawing with a colleague or client and have them add notations for your review using the free mobile app available to smart phones and tablets.

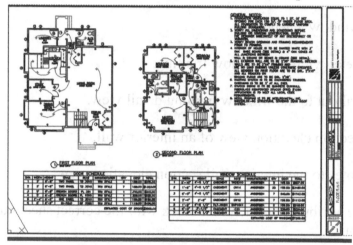

1. When you open the file, you will see a small notification in the lower right corner that a collaborator has added notes to the drawing.

   Click on the link to **Display the Trace palette**.

2. If you click on one of the notes, the drawing will zoom to the area and you will see the notations provided by the collaborator.

3. There is a small toolbar.
   Click on the Green Check.

   You will no longer see the mark-ups.

4. 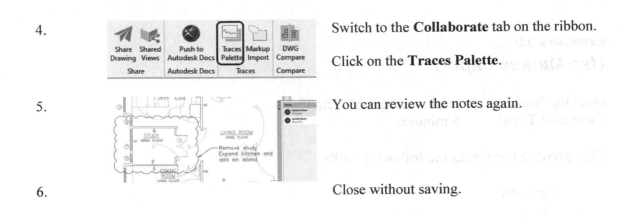 Switch to the **Collaborate** tab on the ribbon.

   Click on the **Traces Palette**.

5. You can review the notes again.

6. Close without saving.

Markup Import

Markup Import attaches an image file or pdf to use as a trace for an existing drawing.

You can convert elements of the trace to AutoCAD elements using the Markup Import tool to make changes quickly and easily.

To create a markup, print out the drawing to be modified. Use a RED pen to indicate the desired changes. Then scan it to a pdf or take a picture with your phone and email it to the person who needs to make the changes.

Exercise 9-35:
Use Markup Import

Drawing Name: markup.dwg, redline.pdf
Estimated Time: 5 minutes

This exercise reinforces the following skills:

- markups
- Trace

1.

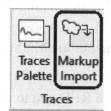

Open the **S1.2 Details** layout tab.

2.

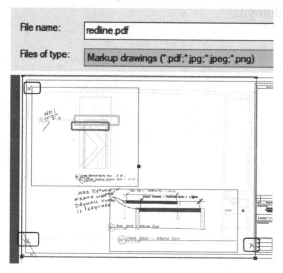

Activate the Collaborate ribbon.

Click **Markup Import**.

3.

File name:	redline.pdf
Files of type:	Markup drawings (*.pdf;*.jpg;*.jpeg;*.png)

Locate the *redline.pdf* file downloaded from the publisher.

Click **Open.**
There are red X marks at three of the corners of the redline pdf to help you align and scale the imported file.

4.

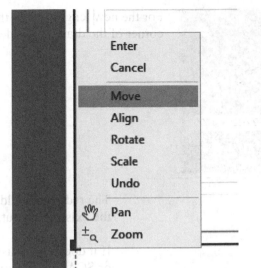

`MARKUPIMPORT Accept placement or [Move Align Rotate Scale Undo] <aCcept>:`

Right click and select **MOVE** or select the MOVE option on the command window.

Select the lower left X on the imported file and align it with the lower left corner of the drawing border.

5.

`MARKUPIMPORT Accept placement or [Move Align Rotate Scale Undo] <aCcept>:`

6.

Click **ENTER** to accept the placement.

Right click in the drawing window.
Select **Basic Modify Tools→Scale**.

7.

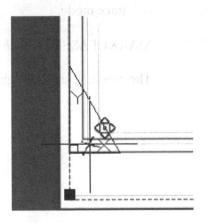

Select the lower left X on the imported file as the basepoint.

8.

`SCALE Specify scale factor or [Copy Reference]:`

9.

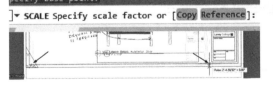

Click on **Reference**.

Select the x on the left side of the imported redline for the start of the reference length. Select the x on the right side of the imported redline for the end of the reference length.

10.

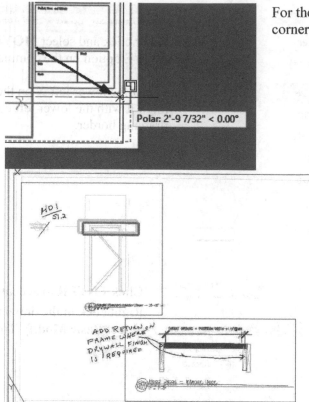

Polar: 2'-9 7/32" < 0.00°

For the new length, select the lower right corner of the drawing border.

The redline should now align with the existing layout.

If it doesn't, try using the MOVE or SCALE tools to adjust it.

It doesn't have to be perfect – just close enough.

The imported PDF is a trace that is placed on top or below the existing drawing.

When you are done with your adjustments it should look like this. Click on the **Edit Trace** icon to exit the edit trace mode.

11.

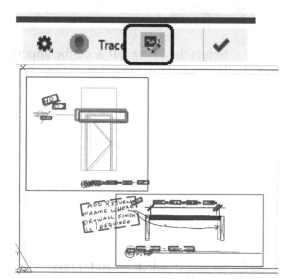

MARKUPASSIST mode is activated.

The markups are highlighted.

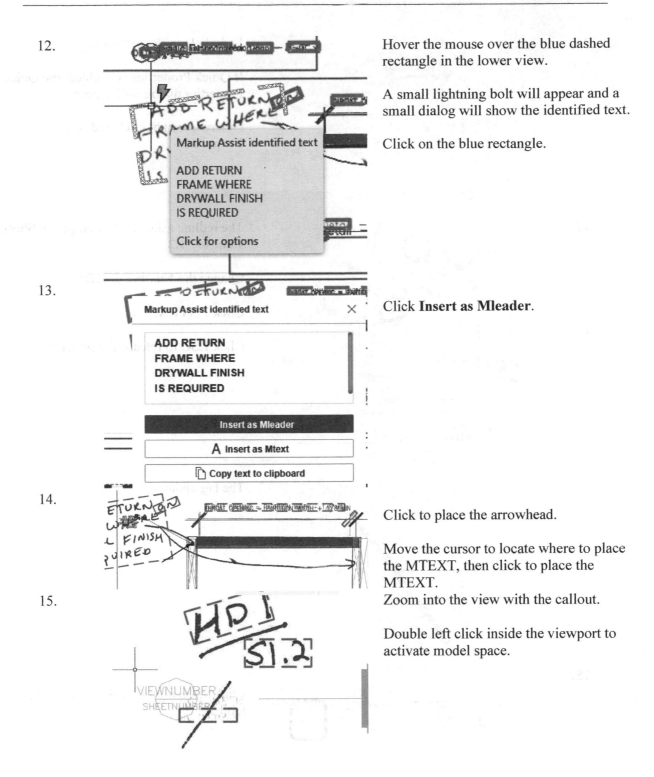

12. Hover the mouse over the blue dashed rectangle in the lower view.

A small lightning bolt will appear and a small dialog will show the identified text.

Click on the blue rectangle.

13. Click **Insert as Mleader**.

14. Click to place the arrowhead.

Move the cursor to locate where to place the MTEXT, then click to place the MTEXT.

15. Zoom into the view with the callout.

Double left click inside the viewport to activate model space.

16.

Left click on the callout tag.

If Quick Properties is enabled, the Quick Properties dialog will appear.

Otherwise, right click and select **Properties**.

17.

The redline asks you to change the Sheet value to **S1.2**.

Change the Number to **HD1**.

Click **ENTER**.

Click **ESC** to release the selection.

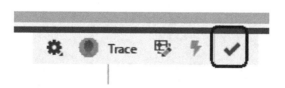

The tag updates.

18.

Click the **Green Check** on the Traces toolbar.

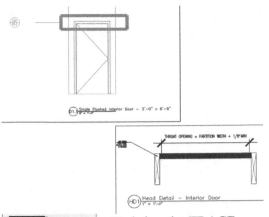

The drawing has been updated.

The imported trace is no longer visible.

19.

To bring the TRACE up again,
Click on the **Traces Palette**.

20.

The trace is still in the drawing.

RMB on the Trace file.

Right click and select **Open**.

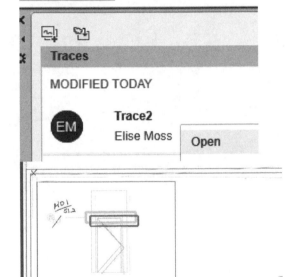

21.

The trace is now visible again.

Green check to close the trace.

Save and close.

Notes:

QUIZ 9

True or False

1. Images cannot be converted to blocks.

2. You can control the size and shape of any elevation view you create.

3. Once you create an elevation view, it cannot be updated.

4. Property set definitions are used in schedules and tags.

5. Property set definitions cannot be used by attributes.

6. You can only place one viewport on a layout.

Multiple Choice

7. You can perform the following in a 2D elevation view: (select all that apply)

   A. Assign a material to individual components
   B. Modify the linework
   C. Render
   D. Change the scale

8. A Live Section View: (select all that apply)

   A. is a 3D view
   B. plan
   C. section
   D. elevation

9. Identify the tool shown.

   A. Expand Window
   B. New Viewport
   C. Rectangle
   D. Paper space

10. If you do not see annotations in a viewport, but they are visible in the Model layout: (select all that apply)

    A. The layer they are on is probably frozen in that viewport.
    B. The annotations require an annotation scale added.
    C. They were deleted.
    D. They are hidden.

11. To add an annotative scale to a tag: (select all that apply)

    A. Select the tag and then use the Add/Delete Scale tool on the ribbon.
    B. Select the tag, right click and select Properties.
    C. Set the tag to be annotative.
    D. Select the tag and use the annotative scale tool on the status bar.

12. To create a custom schedule, you can:

    A. Use a table
    B. Create a schedule style
    C. Use lines and text
    D. Use rectangles and mtext

13. Schedule Table Styles control: (select all the apply:)

    A. Objects that are included in the schedule
    B. Sequence and format of the columns
    C. Property data that can be included
    D. Images of the components

14. Property Data Formats control: (select all the apply:)

    A. Whether a property set definition uses text or numerical data
    B. The units and precision of numerical data
    C. Whether text is upper, lower, or sentence case
    D. The property sets used by different objects, such as doors, windows, or spaces

ANSWERS:

1) F; 2) T; 3) F; 4) T; 5) F; 6) F; 7) A, B, C, D; 8) A; 9) B; 10) B; 11) A; 12) B; 13) A, B, C; 14) A, B, C